Wi-Fi Security

Wi-Fi Security

Stewart S. Miller

McGraw-Hill
New York Chicago San Francisco Lisbon
London Madrid Mexico City Milan New Delhi
San Juan Seoul Singapore Sydney Toronto

The McGraw-Hill Companies

Library of Congress Cataloging-in-Publication Data

Miller, Stewart S.
WiFi security / Stewart S. Miller.
p. cm.
Includes bibliographical reference and index.
ISBN 0-07-141073-2 (alk. paper)
1. Wireless LANs—Security measures. I. Title.

TK5105.78.M55 2003
005.8—dc21

2002043135

1 2 3 4 5 6 7 8 9 0 DOC/DOC 0 9 8 7 6 5 4 3

ISBN 0-07-141073-2

The sponsoring editor for this book was Judy Bass and the production supervisor was Sherri Souffrance. It was set in New Century Schoolbook by Patricia Wallenburg.

Printed and bound by RR Donnelley.

This book is printed on recycled, acid-free paper containing a minimum of 50 percent recycled, de-inked fiber.

B'H

This book is happily dedicated with the greatest love, respect, and admiration to my dear family, who give me the strength and perseverance that truly make life worth living!

CONTENTS

PREFACE

Security is *now* an essential element that forms the cornerstone of every corporate network. Without privacy, however, your solution is *incomplete*! My expertise in the areas of security and privacy has provided me with a valuable perspective that has enabled me to save my clients hundreds of thousands of dollars of what would have been revenue lost to hackers.

Many of my clients ask me to work with their organizations as either a contractor or consultant to assist them in implementing effective security measures because there is no greater cost to an organization than falling prey to a plethora of security vulnerabilities.

As a Director of CyberSecurity for IBM Global Consulting Services for over a decade, I established myself as the leading expert in both network security and enterprise resource planning in several IT sectors. I have published 11 best-selling computer books and have written over 1000 articles for the trade magazines. Today, I am always involved in writing specialized, private analyses for customers interested in acquiring my consulting services.

My experience comes from my extensive work with most of the Fortune 500 companies through my company, Executive Information Services. I work well with personnel to handle the most difficult computing problems, as I am dedicated to creating solutions that specifically meet my clients' individual computing needs.

My core offerings include my work in the following areas:

1. Business and Strategy Development

Project management services that oversee any consulting service through my eyes as an efficiency expert giving you the most "bang for your buck." My expertise is in all areas of security, privacy, and real-world enterprise IT. My company creates the most professional technical writing in white papers, brochures, books, articles, manuals, press releases, and industry "focused" tear sheets.

2. Powerful Market Research
I don't settle for regurgitated market data that puts anyone behind the business curve, and neither should you! In this book and through my company, I provide "real" data, customized to my client's needs to increase both their marketing and sales. I make sure I always stay at the cutting edge of technology by writing comprehensive white papers (technical and/or marketing), market research reports (better than *any* research service), and effectively targeted PowerPoint presentations for greater customer relationship management (CRM). My clients engage me to help them create unique product and business strategies; I then develop that information into computer-based training (CBT) modules to train both staff and clients.

3. Cost-Effective IT Product Selection
Purchasing any IT security product is often prohibitively expensive; my forté is that I save my clients a great deal of money by creating a personalized product comparison matrix to help them get the most functionality for the least amount of money. I cut through the red tape so that my clients should *never* have to overpay for extra costly features they don't require. My work with Fortune 500 companies is so well received because my clients get exactly what they *need*, resulting in savings of several hundred thousand dollars over the long term.

My sole interest in writing this book is to describe the constantly changing face of the wireless world of security and privacy. With information this important, I don't believe anyone should *ever* be treated like a second-class citizen when it comes to getting the right information for their needs and consulting projects.

I have always stuck by one motto: I treat, work, and consider any client's project as though I am working *only* for that one client.

If I can be of assistance in fulfilling your technical writing or analysis needs, please contact me and I will be pleased to assist you in satisfying both your business needs and mission-critical requirements.

Stewart S. Miller
Director, Executive Information Services
Phone: 1-800-IT-Maven
E-Mail: WiFi@ITMaven.com
Web: http://www.ITMaven.com

Wi-Fi Security

CHAPTER 1

Introduction to Wireless LAN Security Standards

Wireless Defined

The wireless industry has evolved phenomenally over the past few years. Wireless transmission (once the domain of amateur radio enthusiasts and the military) is now a commonplace method of data communication for cellular phones, wireless PDAs, text pagers, and, most important, wireless LANs (WLANs).

As there are a number of divergent technologies for wireless networks today (i.e., 802.11b, Bluetooth, etc.) most users standardize on one of these for their corporate networking needs.

The purpose of this chapter is to take a look at the actual security measures that a user must be mindful of in today's business world. There are so many methods and forms of hacker attacks to steal corporate data that wireless measures designed for convenience can be exceedingly harmful without actually taking the proper measures.

Wireless networks are supported by having several transceivers scattered across the typical enterprise to blanket the corporate offices in a web of wireless transmission devices called *access points*. Access points (APs) are strategically placed in fixed locations throughout the company offices to function in tandem like cells of a cell phone network. They function together so that as the computer user moves from office to office, he is still covered by the reception of these wireless network routing devices.

Factors of Security

Primary factors that define security in a wireless environment can be boiled down to five elements; they are shown as tightly integrated interdependent components in Figure 1.1:

1. Theft
2. Access Control
3. Authentication
4. Encryption
5. Safeguards

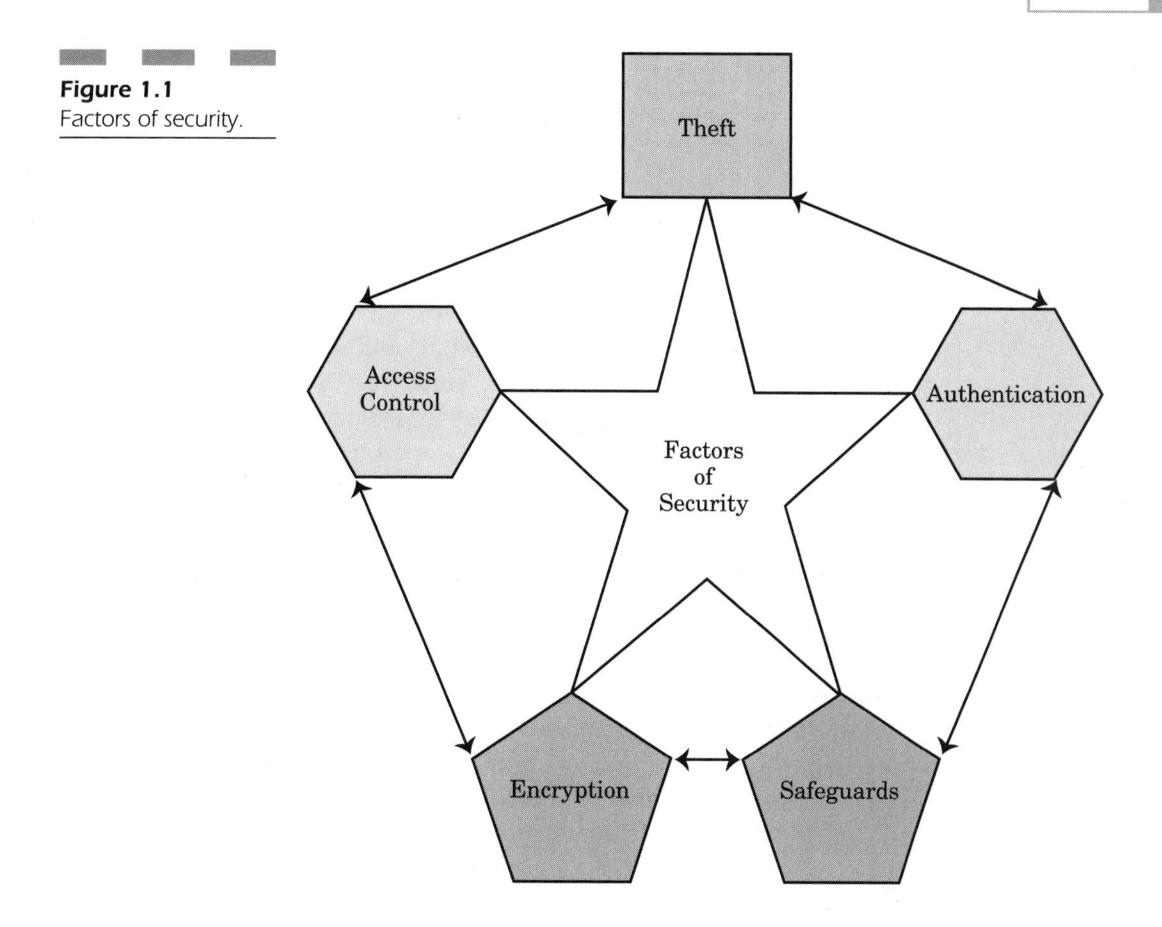

Figure 1.1
Factors of security.

Theft

Unauthorized users often try to log into a network to steal corporate data for profit. Employees who have been terminated often feel resentment and anger against their former employer. It is possible for some users to turn that anger into an attempt to steal corporate data before leaving their company. This is why the easiest type of security measure is simply to disable a user's account at the time of termination. This action is a good security measure and prevents the likelihood of account abuse during the transition out of the company.

Access Control

Many companies set very simple access permissions. You must be wary that networks are designed to increase interoperability so that it is a simple matter for a user to click on his "Network Neighborhood" icon in Windows and see all the wired and wireless devices on his network segment.

Does your company have a policy to set passwords for network shares? Do you know who is sharing what, and with whom? Some users want to share a document from one employee to another so they just share "Drive C" on the network. But if they don't remember to turn off the sharing, everyone within that network segment has full read and write permission to that user's Drive C. If there is a virus running across your network, you can bet that that user's computer is fully vulnerable and will most definitely be compromised.

Wireless networks not only have all the same access control vulnerabilities as wired networks, but they can easily be accessed by outsiders. The most common type of attack is simply to sit outside an office building and use a wireless network interface card to roam onto any available 802.11b network. Since the majority of users fail to set even the simplest access control barriers that prevent a random user from accessing the network, everything on your network becomes vulnerable to attack, theft, or destruction from a virus.

Authentication

Do you know if the user logged in is really that person? It is an all too common practice for people to use other people's accounts to authenticate themselves to the server. In most wireless networks, businesses often configure one account, "Wireless User," and that account can be used by several different devices. The problem is that a hacker (with his own wireless device) could easily log onto to this general account and gain access to your network.

To prevent an unauthorized user from authenticating himself into your network, you can set your router to permit only connections from authorized wireless network cards. Each wireless network card has a Media Access Control (MAC) address that uniquely identifies it. You can tell your router only to authenticate those wireless users with a network card that is pre-authenticated to use your network. This protects you against users who are trying to gain access to your system by roaming around the perimeter of your building looking for good reception to log onto your local area network.

Encryption

If a user is not able directly to log into your network, he may use a wireless "packet sniffer" to try and eavesdrop on the network traffic. In that way, even if the hacker is unable to authenticate himself onto your network, he can still steal sensitive corporate data by monitoring your traffic for usable information. In addition to viewing private data files, the hacker is potentially able to "sniff" usernames, passwords, and other private information to gain access onto your network.

Wireless routers support medium and strong levels of encryption that scramble the data and make it unusable to anyone trying to eavesdrop on the network traffic. Only the users at either end of the "authorized" connection can view and use the data.

Unfortunately, most users don't turn on encryption in their wireless devices to protect themselves against eavesdropping! Most wireless routers have an internal Web site that allows for the very simple and easy configuration of data privacy. Wired equivalent privacy (WEP) is a security protocol for wireless local area networks (WLANs) designated by the 802.11b standard. WEP offers a level of security similar to that of a wired LAN.

Wired LANs offer greater security than WLANs because LANs offer the protection of being physically located in a building, whereas a wireless network inside a building cannot necessarily be protected from unauthorized access when no encryption is used. WLANs do not have the same physical confinements and are more vulnerable to hackers. WEP provides security by encrypting data over radio waves so that it is protected as it is transmitted from one end point to another. WEP, used on both data link and physical layers, does not provide point-to-point security.

Most wireless routers offer 64- and 128-bit encryption with a user-specified encryption key that scrambles your data according to your input. This key is needed at points to decode the data into a usable form. Most users, however, keep this option disabled and therefore are vulnerable to anyone intercepting network traffic or even roaming onto the network.

Safeguards

The best safeguard is to become familiar with your WLAN and your wireless router. You should take the steps above into serious consideration and establish an encryption key at least at the 64-bit, but preferably at the 128-bit level. However, it is important to note that some wireless network cards only support the lower level of encryption; many companies often charge a few more dollars to have their cards support the 128-bit encryption scheme.

Once you have turned on encryption, you may then see that your system supports traffic only by wireless network card MAC addresses that you can specify. This precludes someone's trying to break into your WLAN from outside your building or from the parking lot that is in range of your wireless transceiver array.

As a network administrator, there are a number of ways you can safeguard your WLAN against intrusion by following some very simple, commonsense steps to make certain you are not being hacked. Wireless routers always have an activity light that shows you when traffic is flowing across the WLAN. There are also a number of software utilities that measure network traffic, where that traffic is going, and the throughput of each connection (how fast a download is proceeding).

If you see an unusual amount of network traffic flowing across your wireless network and the activity light of your wireless router is congested with an enormous amount of traffic, then you know something is wrong! You can trace each connection into the router and if there is a connection that doesn't belong then you know someone may have hacked into your system. Commonsense types of safeguards would indicate that a normal user wouldn't be using the wireless connection to capacity for any prolonged period of time. Those types of connections are established for the purpose of drawing out information, databases, and files from your network for corporate espionage.

Sometimes just realizing that your WLAN can penetrate the walls of your office, building, and workgroup is enough to help you realize that it is easily possible for someone to try and break into your system from anywhere on your immediate perimeter. Just make certain you account for all the network traffic; it is also a good idea to keep a log of all network activity. If someone does try to hack into your network, there will be a tremendous spike in activity during different periods of the day or night. You can then use that log to isolate unusual network activity and place safeguards on your network to keep an eye out for suspicious activity. This type of safeguard is akin to an "intrusion detection system," which alerts you to fraudulent and unauthorized access attempts into your network from any external source.

Intrusion Detection Systems

Since I am pointing out some important safeguards for your WLAN, this is the place for a brief introduction to the intrusion detection system. There are a number of commercial solutions that use rules-based technology to determine "automatically" if someone is trying to hack your wireless network, while other have "real" human beings study your logs for suspicious activity.

An intrusion detection system (IDS) checks out all inbound and outbound network activity and identifies any suspicious types of activity that indicate a network or system attack from a hacker trying to breach your WLAN.

Primary types of IDS, as shown in Figure 1.2, include:

- **Pattern detection**—An IDS analyzes the information it collects and compares it to large databases of attack signatures. The IDS looks for a specific attack pattern that has already been documented. This type of detection software is only as good as the database of hacker attack signatures that it uses to compare packets to. The system administrator can also designate anomalies that stray from the normal network's traffic load, breakdown, protocol, and typical packet size. The IDS monitor detects network segments to compare their state to the normal baseline and looks for anomalies that match a specified pattern of attack.
- **NIDS and HIDS**—Network- and host-based intrusion detection system analyze individual packets flowing through a network. NIDS can detect malicious packets that get past your firewall filtering rules. Host-based systems examine the activity on each individual computer or host.
- **Passive and reactive systems**—The passive system IDS detects a potential security breach, logs the information, and sends an alert. The reactive-system IDS responds to the suspicious activity by logging off a user or by reprogramming the firewall to block network traffic from the suspected hacker.

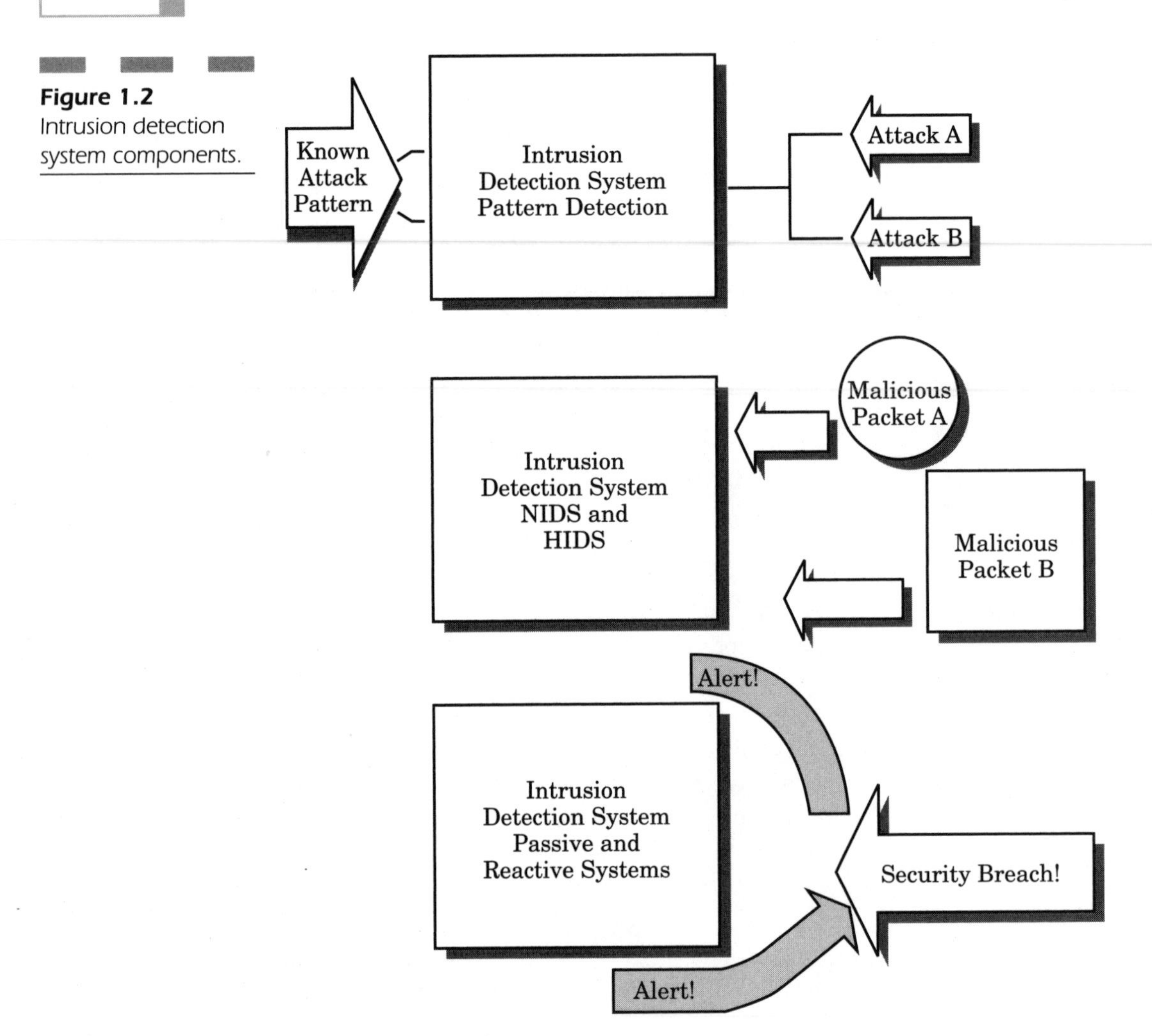

Figure 1.2 Intrusion detection system components.

Each IDS differs from a firewall in that a firewall looks out for intrusions in order to stop attacks from occurring. The firewall restricts the access between networks in order to stop an intrusion; however, it does not usually catch an attack from inside the network. An IDS, however, examines the suspected intrusion once it has taken place and sends an alert. Note than an IDS also looks for attacks that originate from within a system. This can easily occur when a wireless network user appears to be an "internal user" of your wireless network and therefore hard to distinguish from a legitimate user.

IEEE

The Institute of Electrical and Electronics Engineers (IEEE) is an organization composed of engineers, scientists, and students. It has developed standards for the computer and electronics industry. The focus here is on IEEE 802 standards for wireless local-area networks.

WECA

The Wireless Ethernet Compatibility Alliance is an organization composed of leading wireless equipment and software providers with the mission of guaranteeing interoperability of Wi-Fi products and to promote Wi-Fi as the global wireless LAN standard across all markets.

Wi-Fi

Wi-Fi is an acronym for wireless fidelity, commonly seen as IEEE 802.11b. The term comes from WECA. Wi-Fi is synonymous with 802.11b in much the same way as Ethernet is used in place of IEEE 802.3. Products certified as Wi-Fi by WECA are interoperable regardless of manufacturer. A user with a Wi-Fi product can use any brand of access point with any other brand of client hardware that is built to use Wi-Fi.

The Many Flavors of 802.11

The 802.11 standard is defined through several specifications of WLANs. It defines an over-the-air interface between a wireless client and a base station or between two wireless clients (Figure 1.3).

There are several specifications in the 802.11 family:

- **802.11**—Pertains to wireless LANs and provides 1- or 2-Mbps transmission in the 2.4-GHz band using either frequency-hopping spread spectrum (FHSS) or direct-sequence spread spectrum (DSSS).

- **802.11a**—An extension to 802.11 that pertains to wireless LANs and goes as fast as 54 Mbps in the 5-GHz band. 802.11a employs the orthogonal frequency division multiplexing (OFDM) encoding scheme as opposed to either FHSS or DSSS.
- **802.11b**—The 802.11 high rate Wi-Fi is an extension to 802.11 that pertains to wireless LANs and yields a connection as fast as 11 Mbps transmission (with a fallback to 5.5, 2, and 1 Mbps depending on strength of signal) in the 2.4-GHz band. The 802.11b specification uses only DSSS. Note that 802.11b was actually an amendment to the original 802.11 standard added in 1999 to permit wireless functionality to be analogous to hard-wired Ethernet connections.
- **802.11g**—Pertains to wireless LANs and provides 20+ Mbps in the 2.4-GHz band.

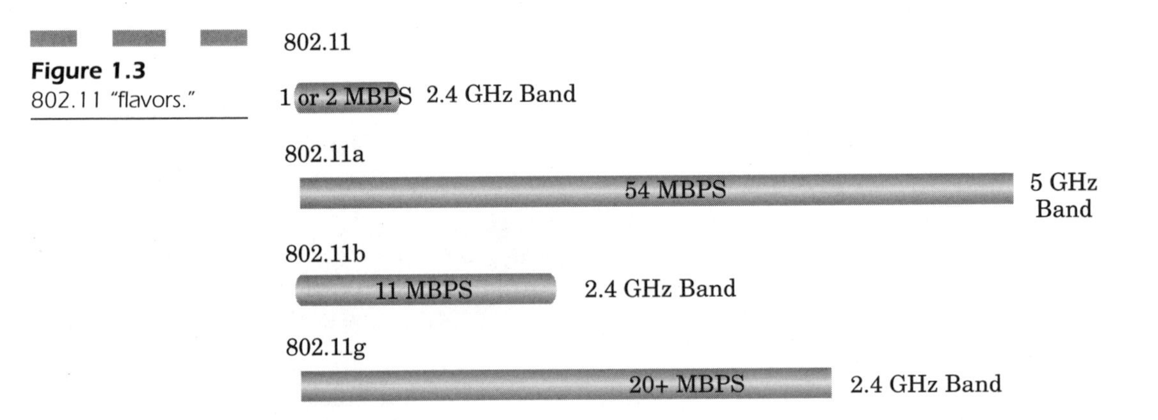

Figure 1.3 802.11 "flavors."

FHSS

FHSS is an acronym for frequency-hopping spread spectrum. There are two types of spread spectrum radio, FHSS and direct-sequence spread spectrum. FHSS is a transmission technology used in local area wireless network (LAWN) transmissions where the data signal is modulated with a narrowband carrier signal that literally hops in a random but predictable sequence from frequency to frequency. The calculation is a function of time over a wide band of frequencies. The signal energy is spread in time domain rather than slicing each element into small pieces in the frequency domain. This technique reduces interference because a signal from a narrowband system will only affect the spread-spectrum signal if both are transmitting at the same frequency at the same time. If this is synchronized correctly, just one logical channel is supported.

The transmission frequencies are designated by a *spreading code*. The receiver must be configured to the same spreading code and must listen to the incoming signal at the correct time and frequency in order to receive the signal properly. Federal Communication Commission regulations require manufacturers to use 75 or more frequencies per transmission channel with a maximum time spent at a specific frequency during any single spread at 400 ms.

DSSS

DSSS is an acronym for direct-sequence spread spectrum, which employs frequency-spreading spread spectrum. DSSS is a transmission

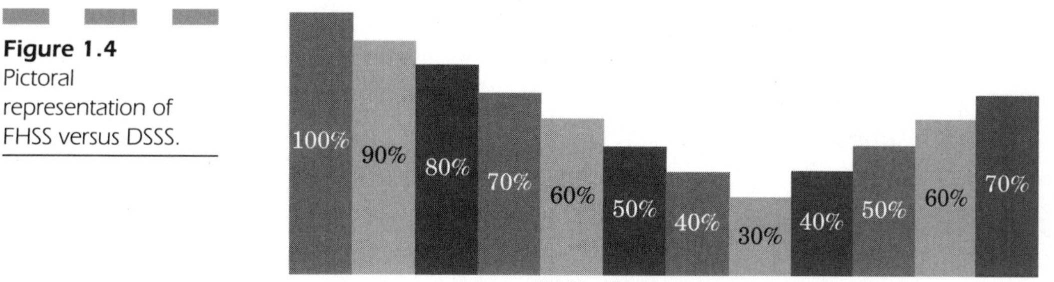

Frequency Hopping Spread Spectrum

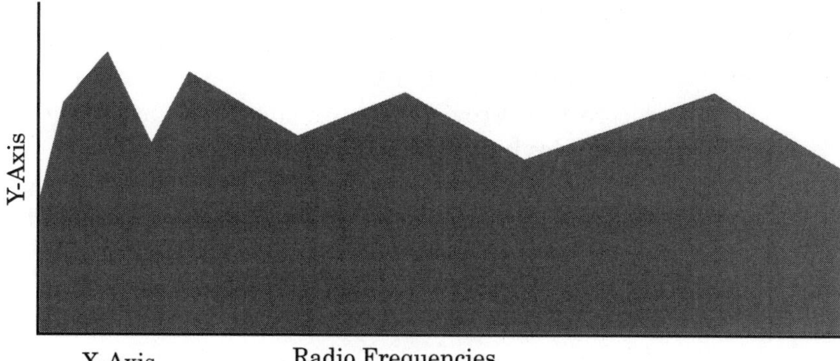

Direct Sequence Spread Spectrum

Figure 1.4 Pictoral representation of FHSS versus DSSS.

technology used in LAWN transmissions where a data signal at the sending station is joined with a higher data rate bit sequence or *chipping code*, which slices the user data according to a spreading ratio. The chipping code is a redundant bit pattern for each bit transmitted. This increases the signal's resistance to interference. Should one or more bits in the pattern be damaged during transmission, the original data can be recovered because of the amount of increased redundancy of the transmission.

OFDM

OFDM is an acronym for orthogonal frequency division multiplexing, an FDM modulation technique for transmitting large amounts of digital data over a radio wave. OFDM works by slicing the radio signal into multiple smaller subsignals that are then transmitted to the receiver at the same time over different frequencies. OFDM reduces crosstalk in signal transmissions. 802.11a WLAN technology uses OFDM.

Bluetooth

Bluetooth is a short-range radio technology that creates a simpler method of communicating across networked devices and between devices and the Internet. It also simplifies data synchronization between net devices and other computers (Figure 1.5).

Products with Bluetooth technology must be qualified and pass interoperability testing by the Bluetooth Special Interest Group prior to being released on the market.

The Bluetooth 1.0 specification consists of two parts: a foundation core that provides design specifications and a foundation profile that provides interoperability guidelines.

Bluetooth was created in part through the cooperation of several companies including Ericsson, IBM, Intel, Nokia, and Toshiba.

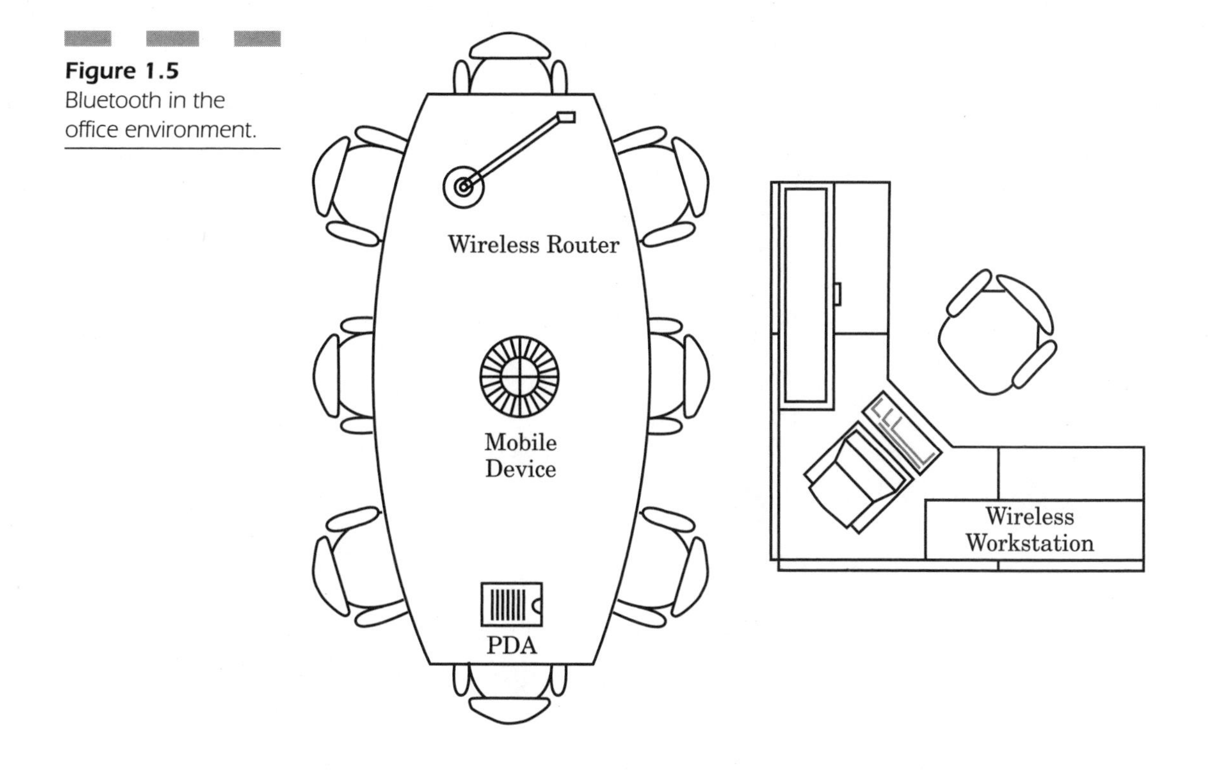

Figure 1.5 Bluetooth in the office environment.

Differences between the Wireless Standards

The 802.11b standard is more common than Bluetooth. In fact, Windows XP operating system supports many WLAN NIC cards by default. More and more cards are supported under Linux, Windows CE, and Pocket PC.

Macintosh computers running either System 9 or OS X have their own version of 802.11b, called "airport" cards. These cards are simple 802.11b cards that function in tandem with any wireless router or other similarly equipped PC on a WLAN. There are a number of utilities (i.e., DAVE by Thursby Software) that make the Mac computer look just like a Windows workstation on a generic wireless LAN. You can wirelessly transfer files, surf the Internet, or log onto any number of wireless domain servers in your corporate offices.

The maximum speed of 802.11b is 11 Mbps, but that speed is dependent upon your proximity to the wireless router or transmitter. As you increase your distance from the wireless transmitter your speed decreases to as low as 2 Mbps at maximum distance.

There are several factors that control the range of your wireless transmitter. When you are outdoors, you have better reception because there aren't any items that block your signal. Indoors, you have to contend with building materials, shielding in the walls, and other equipment that can generate electrical interference that can disrupt or corrupt your wireless signal.

Bluetooth has a maximum speed of 2 Mbps and suffers the same limitations in its radio frequency interference pattern as 802.11b. Bluetooth is a competing standard that is currently being built into mobile phones, PDAs, and network interface cards for PCs. This standard is supposed to have more far-reaching implications as it is to be adopted in more devices. However, because its maximum speed is lower than that of 802.11b (and other 802.11 standards) it does not have the same far-reaching implications for higher-speed wireless networks.

Conclusion: How Security Applies

In dealing with radio signals, you must be wary that you no longer have the "security" or physicality of a hardwired line. When cellular phones came out, the biggest problem was that people who had scanners could listen into private conversations. This made wired phones essential for private communication. In order to tap a wired phone, or a wired LAN for that matter, you would have to have a packet sniffer directly attached to the wire listening to the network traffic.

Wireless networks can be "sniffed" from any portable computer with a wireless networking card. This is why encryption is so important. If someone is interested in listening into your private network traffic, you should at the very least make it extremely hard for them to decode your transmission. Most hackers won't keep trying if they can move onto an easier target.

You should note that you cannot rely entirely on wireless encryption methods because they can be compromised given a reasonable amount of time. If you are concerned about security, use the highest-strength encryption available to your system (usually 128-bit). Make it a point to change the encryption key as often as possible (at least once every week or two) just to make it difficult for someone "sniffing" your wireless network in an effort to decode your encryption key and log onto your WLAN to steal, corrupt, or damage your mission-critical data.

Finally, set your router to accept only incoming connections from wireless network cards that you trust within your organization. Don't leave yourself vulnerable to hackers trying out a "parking lot attack" on your system. This is when someone sits outside your building (in a car) either on the street or next to a window right on the fringe of reception. Hackers then attempt to compromise your systems by logging into your WLAN as though they were an actual employee within the confines of your building. If you tell your router to screen out unknown network cards (each card has its own unique identifier called a MAC address) then you add at least another layer of protection to help keep your network isolated from security breaches so your WLAN won't get hacked!

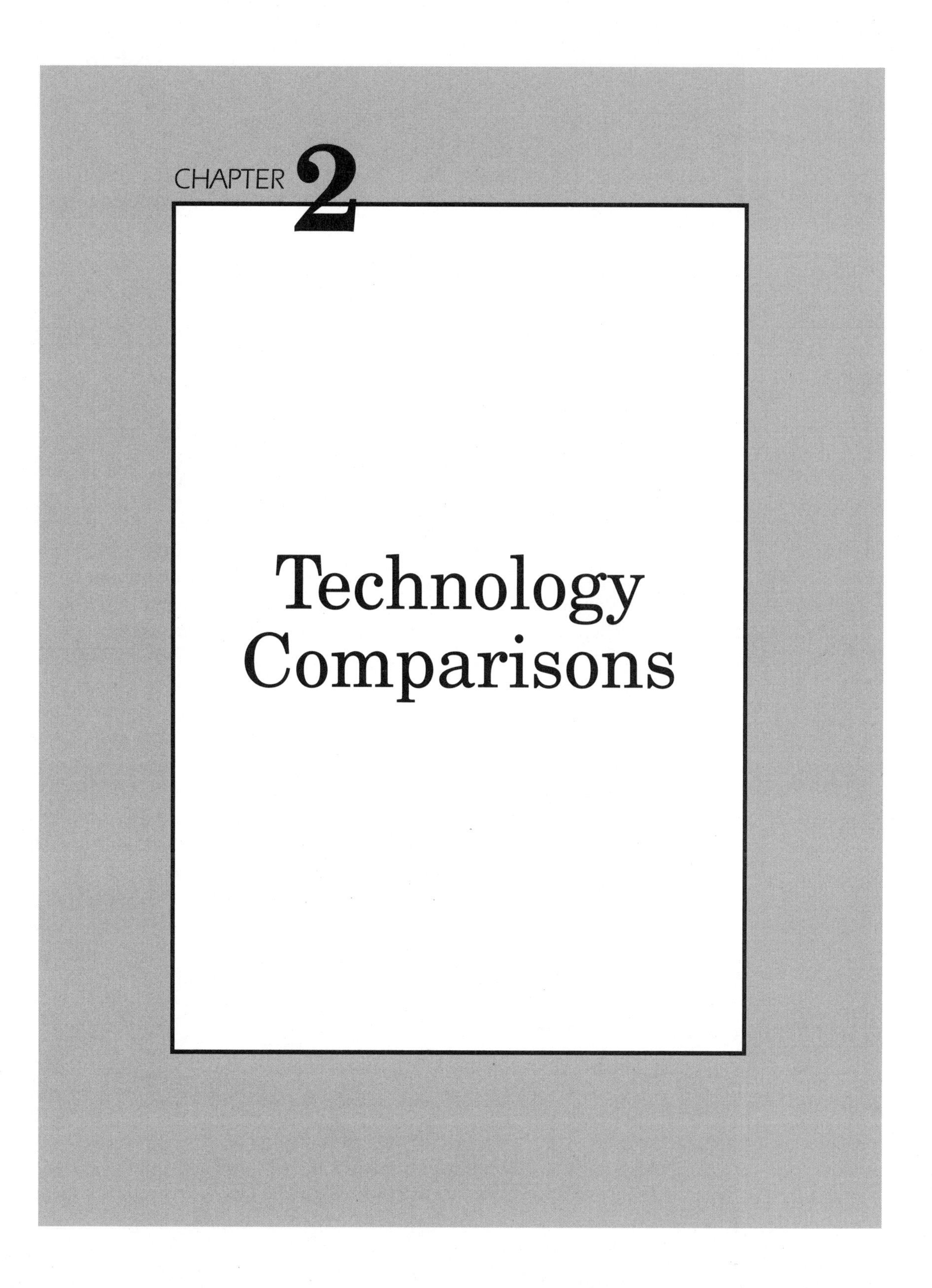

CHAPTER 2

Technology Comparisons

This chapter looks at how 802.11b stacks up against other wireless standards and specifications. From a security standpoint, it is essential to understand the literal nuts and bolts of the 802.11 standard in comparison to both Bluetooth, HomeRF, and SWAP. We see how all three protocols offer security measures to protect data and privacy, but we also see exactly how effective these measures are and the type of controversy they experience when used for mission-critical business applications.

HomeRF

HomeRF is a collaboration of several big companies from varied backgrounds to design a form of wireless LAN (WLAN) that functions in both the home and small-office environment. This group also is working towards the development of the SWAP LAN standard.

WLANs are becoming increasingly popular in home and home office environments much the same way that cordless phones have come to be integrated into our lifestyle for practically every application. The home market represents ideal territory because most homes are not built with LAN cabling and it becomes essential to transport computer resources from one room to another.

HomeRF's main concern has been to deploy itself cost effectively in a WLAN. Since cost is still a limiting factor over wired LANs, most wireless users cannot justify spending the money to purchase wireless network interfaces cards or a wireless routing access point device. The reason more and more people are buying 802.11b is that prices have dropped considerably in 2002, making wireless NIC cards and access points much more cost effective. Since so many vendors are selling 802.11b, there is a higher degree of competition and pressure to keep costs competitive, whereas HomeRF really hasn't taken off as much as the 802.11 standard has.

802.11 versus SWAP

The 802.11 specification was designed to have more restrictive timing and filtering patterns as opposed to SWAP, which did not tightly adhere to these regulations and was therefore easier to implement at lower costs.

Note that MAC is implemented both in the software and digital layers and doesn't really factor into the costs involved. SWAP relaxed some of those hardware constraints in an attempt to make the medium less complex, but with fewer features and less functionality than its 802.11 counterpart.

SWAP Specification

The SWAP specification is an open standard, even more so when compared to 802.11, as there are no royalty or patent issues to contend with. The specification is simple and permits the combination of both voice and data.

Integrating Wireless Phone and Data

A redesign of the MAC protocol has offered the integration of the more refined features of DECT (an ETSI digital cordless phone standard) and the 802.11 standard. The idea is the creation of a digital cordless phone on an ad hoc data network.

This device can carry voice services over TDMA protocol while implementing protection mechanisms to avoid interference patterns. The concept uses DECT in combination with a voice codec. The data elements use CSMA/CA access, which is analogous to 802.11 along with MAC retransmissions and fragmentation mechanisms, to provide an environment parallel to a wired Ethernet.

If this does become a reality then there could easily be a 1–2-Mbps frequency-hopping physical layer that permits as many as six simultaneous voice connections while still providing sufficient data throughput for regular users. In addition, the voice quality will be at least as good as, if not better than, current digital phone implementations. However, data throughput may be less when compared to 802.11b.

Bluetooth

Bluetooth is often compared to 802.11, but is really distinctly different. It was conceived as a wireless replacement for a wired Ethernet and was developed by Ericsson with the assistance of Intel.

Bluetooth provides point-to-point links without any native IP support, meaning it cannot easily support point to point protocol (PPP). You *can* create of a set of point-to-point wireless serial conduits, referred to as RfComm, between the master machines and as many as six slave machines using the session definiation protocol (SDP) to bind those conduits to a specific driver or application.

Nodes must be explicitly connected, however they do recall bindings each time they are used. Bluetooth does support TCP/IP as one profile implemented through PPP on a given conduit. Additionally, there are conduits for audio and other wireless applications as well.

The difficulty when comparing Bluetooth to any WLAN is that it does not truly support applications including:

- Native IP support
- Cellular deployment
- Connectionless broadcast interface

The most fundamental drawback is that Bluetooth doesn't combine TCP/IP and WLAN applications nearly as well as 802.11 does. In contrast, Bluetooth is a good implementation for such applications as a wireless universal serial bus (USB), something 802.11 has not been able to accomplish as easily. This is because TCP/IP discovery mechanisms and binding protocols can't support wireless USB applications as well.

Wireless Hacking

As the technology for wireless applications has broadened to encompass simple networking at airports, banking, and financial institutions, wireless security is a main concern since these mechanisms could inadvertently be broadcasting your personal information to other wireless sniffing devices.

For example, there are specific products that can help you eavesdrop on wireless transmissions.

NetStumbler

NetStumbler is a utility that works under Windows and is meant specifically for 802.11b wireless networks. Under the 802.11 standard, most wireless hardware vendors are compatible. However, security is the most commonly overlooked element. NetStumbler has released its software in an effort to increase awareness of the inherent problems with wireless communication security, or lack thereof. The objective is to see that vendors concentrate on security while maintaining the functionality necessary in wireless products. An example of a wireless hacker is shown in Figure 2.1.

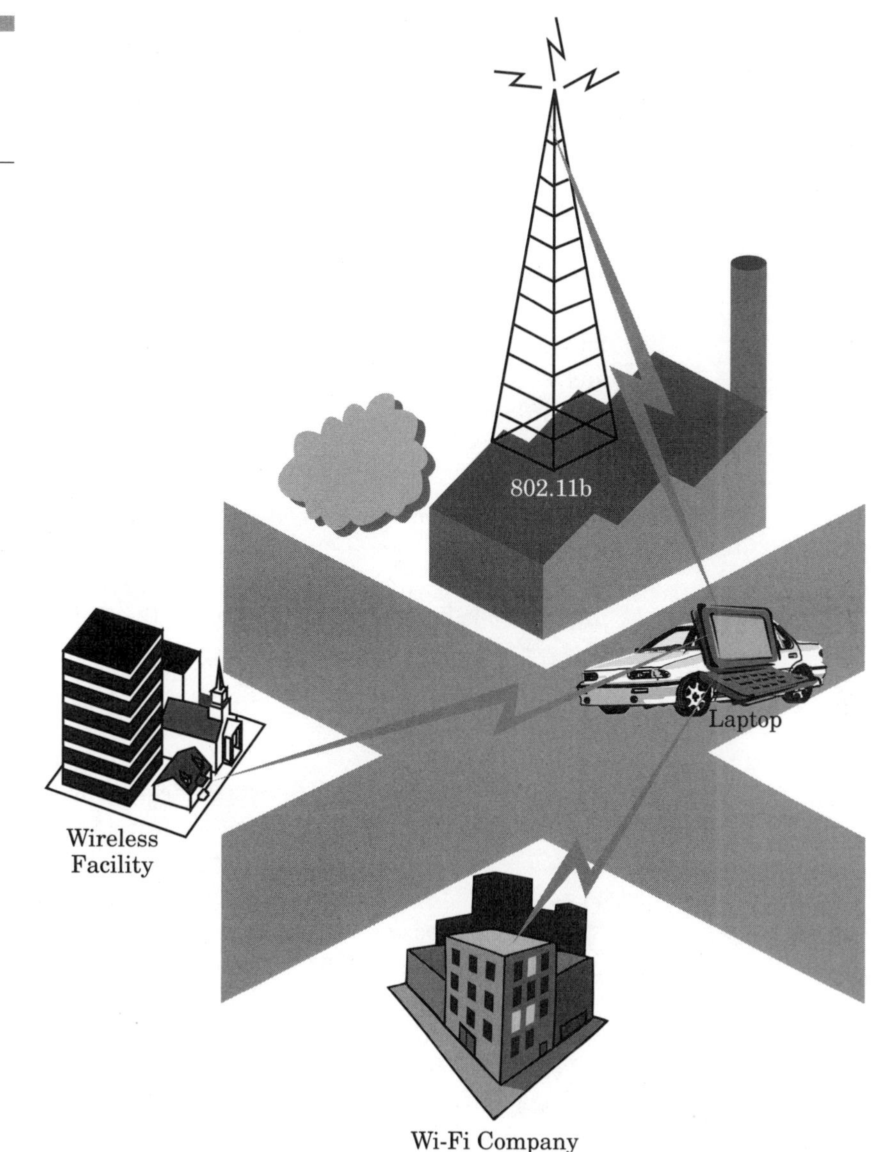

Figure 2.1 Wireless hacker looking to "stumble" on a Wi-Fi signal.

NetStumbler Software Uses

The actual software that NetStumbler produces is primarily used by security consultants who find it a necessity to check that their corporate WLANs are not wide open to the public. It is all too common for hackers to access wireless networks from the street or outside corporate facilities and consume your precious wireless network bandwidth.

A system administrator can use this tool in order to check how far the wireless coverage extends from the WLAN into surrounding areas. NetStumbler allows the collection of demographics information about the 802.11 network. This information can be used to prevent hackers from accessing the network from nearby offsite locations, or keeps any overly curious person from accessing a network.

This type of tool is most useful in determining what types of wireless network exist in almost any location. Unsecured wireless networks exist in more public places that you might think (Figure 2.2), including:

- Public office buildings
- Malls
- Tax preparer offices
- Restaurants

The benefit and problem of setting up wireless LANs is that it is all too easy for companies or small offices to supply employees with 802.11b devices. All that is really necessary is to set up an access point at the wired LAN server, and everybody is connected on the wireless corporate network.

Script Kiddies

Hacking is no longer limited to experienced or malicious users who are trying very hard to roam onto your corporate network, steal your information, and access your network resources at no cost. Today, hacking is just as accessible to teenagers who need only to understand how to execute a program, access a resource, or just launch a program. These people are commonly referred to as *script kiddies*.

Script kiddies are people who are more curious than malicious; they want to see how many and what kinds of wireless resources are accessible to them. Wireless LANs are like candy to these users. You don't even have to be "inside" the actual company walls in order to be able to access 802.11b resources; that is what makes hacking a WLAN so appealing.

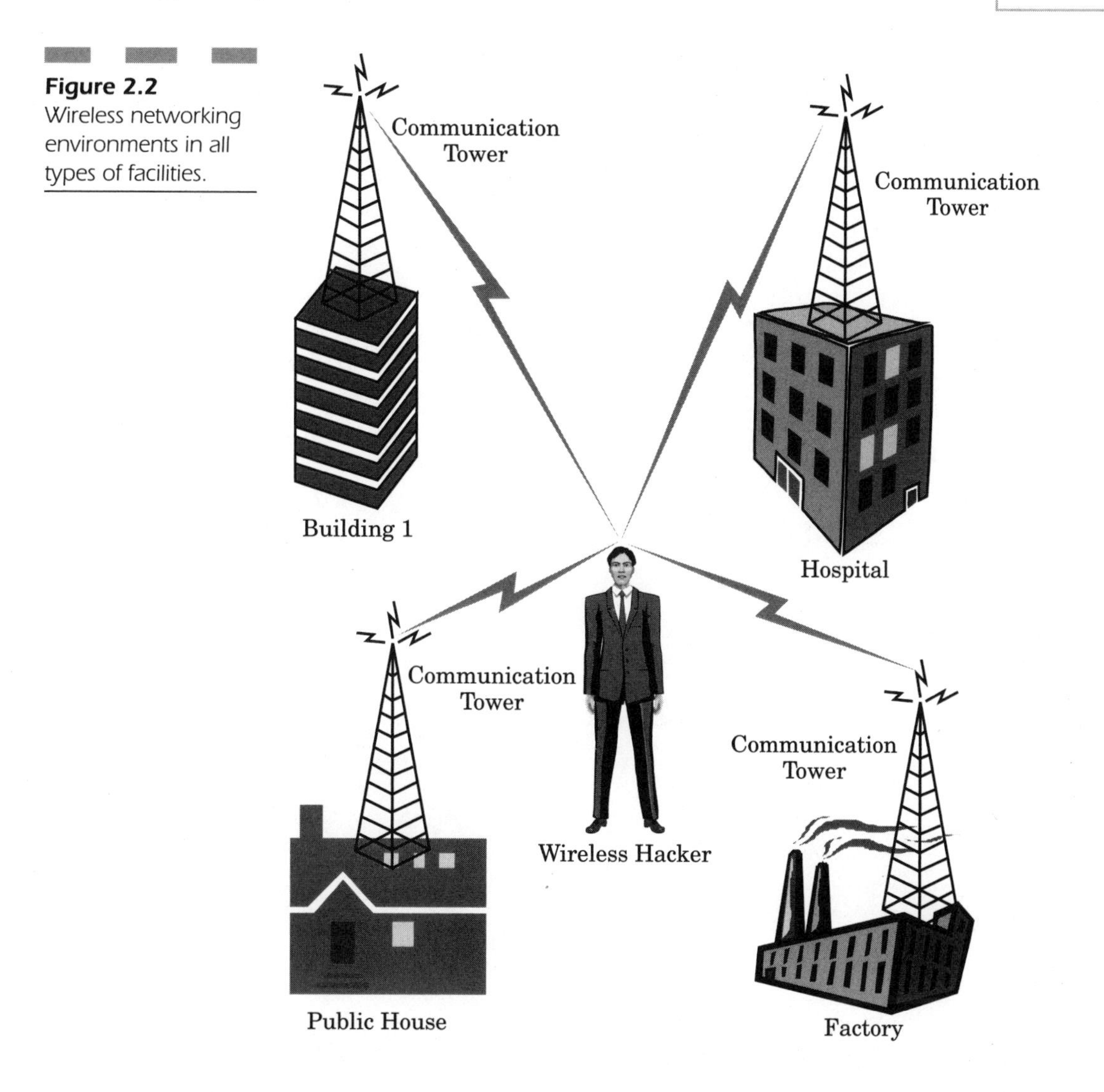

Figure 2.2 Wireless networking environments in all types of facilities.

A cartoon recently published in the security journals depicted a long-haired computer user sitting in his car on a public street, driving through a neighborhood. The driver stops outside one house and starts using his wirelessly enabled laptop because he found an 802.11b open system that was accessible while sitting inside his car. The homeowner realizes his 802.11b network has been compromised and looks outside his window and notices the individual. He walks out to confront the man in the car who quickly replies, "I'm sitting here on a public street with my own equipment, it's not my fault that your signal is leaking out into the public street!"

This is how many script kiddies justify using network resources. Since these resources are accessible from distant locations, accessing free Internet resources almost begs script kiddies to try to gain access.

Facts

As the technology for 802.11 has become both popular and inexpensive, a number of common problems have resulted from users' inability to provide effective security measures (Figure 2.3). For example:

- Administrators don't create a unique service set identifier (SSID)—an identifier that determines a specific network
- Only a quarter of corporate WLANs use wired equivalent privacy (WEP)—an 802.11 standard form for encrypting traffic

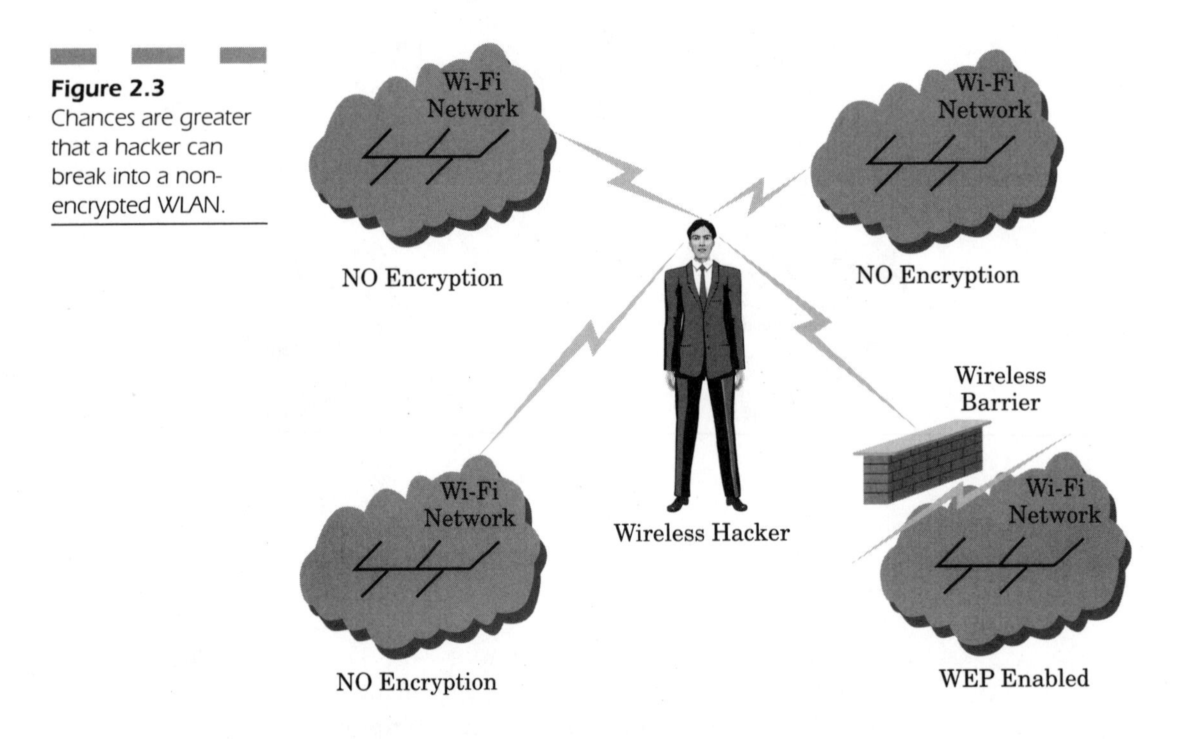

Figure 2.3 Chances are greater that a hacker can break into a non-encrypted WLAN.

Even though you hear about multiple vulnerabilities in using 802.11 with WEP, there are still significant benefits in using it as opposed to running an open system network.

Bluetooth Technology

Bluetooth technology represents a globally used open and short-range radio specification that concentrates on the communication between the Internet and networked devices. In addition, it designates communication protocols between both devices and computers.

Bluetooth certification is achieved by devices that pass interoperability testing by the Bluetooth Special Internet Group (SIG), an entity which makes certain that these products satisfy the standard.

Bluetooth is based around a 9mm by 9mm microchip that operates as an inexpensive means of forming a short-range radio link that provides security for fixed wireless workstations as well as mobile computing devices such as PDAs. All this effort is designed to eliminate the massive amount of cables and connections that link every device we require in the modern computing environment of the office.

One of the most advantageous features that Bluetooth has to offer is that it can network devices "ad hoc." This means you can link your laptop computer, PDA, and phone with one centralized Bluetooth interface. You can transfer files, names, and addresses with one unified connection protocol, essentially breaking the barrier of sharing information from one device to another.

Bluetooth Background

Bluetooth was originally formed by the following five entities: IBM, Intel, Ericsson, Nokia, and Toshiba.

The initial five have grown to well over a thousand companies at this point and the number is increasing. Though Bluetooth is still not quite as popular as 802.11, there are a number of real potential applications for a wide array of divergent wireless devices.

For interesting background, where exactly did Bluetooth get its name? Contrary to what your dentist might think, it is not from eating blueberries. The actual origin of this term is from a 10th-century Scandinavian king whose name was Harald Bluetooth. The connection is that in his real life, he managed to unite several disparate kingdoms under one area. The idea was to make Bluetooth encompass a kingdom of different devices and to create a convergence of many different devices under the umbrella of one global specification.

What Gives Bluetooth Its Bite?

The Bluetooth technology has specific features and functionality that gives it the ability to encompass a divergent set of varied technologies. These features include:

- Bluetooth segments the frequency band into hops. Spread spectrum is used to hop from one channel to another. The result is the addition of a stronger security layer.
- Up to eight devices can be networked in a personal area network (PAN), a conglomerate of devices connected in an ad hoc way using Bluetooth. A piconet or PAN is formed when at least two devices (i.e., a portable PC and a cellular phone) connect. A PAN can support as many as eight devices. When a PAN is formed, one device acts as the master while the others act as slaves for the duration of the connection.
- Signals can be transmitted through walls and containers, eliminating the need for line of sight.
- Devices do not need to be pointed at each other, as signals are omnidirectional.
- Both synchronous and asynchronous applications are supported; this simplifies the process of implementing Bluetooth on a variety of devices and for a variety of services including both voice and Internet.
- Regulation by governmental agencies makes it easy to implement a wide array of implementations globally.

Bluetooth defined Since the terminology for this technology can be quite intensive, it is important first to learn the definitions for the most common Bluetooth terms:

- **Piconet or personal area network (PAN)**—Devices connected in an ad hoc manner that don't need to be predefined or planned in advance (as opposed to a wired Ethernet). As few as two or as many as eight devices can be networked into a piconet. This represents a peer network, meaning that when it is connected, each device has equal access to the others. Note: One device in the chain is the master, while the others are slaves.
- **Scatternet**—Several piconets or PANs can form a larger scatternet, where each piconet is completely independent.
- **Master unit**—The master in a piconet or PAN whose clock and hopping sequence synchronizes all the other devices.

- **Slave unit**—Devices in a piconet or PAN that are not the masters.
- **MAC address**—Three-bit address that uniquely identifies each unit in a piconet or PAN.
- **Parked units**—Piconet devices that are synchronized, though they don't have MAC addresses.
- **Sniff and hold mode**—Power-saving mode of a piconet device.

Bluetooth topology The typical Bluetooth network topology is either point to point (P2P) or multipoint.

A PAN can establish a connection to another PAN to form a "scatternet." The typical scatternet has four units connected to a PAN that has two units. It is important to note that the "master unit" in this scheme is *not* the actual connection link between the two PANs.

Transmission speed Both circuit and packet switching combine within the baseband protocol, which involves one single channel per line. One major concern is that packets do not arrive out of sequence. To avoid these problems, as many as five slots can be reserved for synchronous packets. Note that a different hop signal is used for each individual packet.

Using a baseband protocol, circuit switching can be either synchronous or asynchronous.

You can have as many as three synchronous voice or data channels, such that one synchronous and one asynchronous data channel can be supported on any given channel. This means that each synchronous channel can support as much as 64-Kbps transfer speeds (sufficient for voice transmissions).

In contrast, asynchronous channels can send as much as 721 Kbps in one direction with as much as 57.6 Kbps in the opposing direction. Furthermore, you can also have an asynchronous connection supporting 432.6 Kbps in both directions when you have symmetric link.

Bluetooth Spectrum Hopping

One of the advantages of Bluetooth, like 802.11, is that it uses frequency hopping and fast acknowledgment that result in enhancing the connection and isolating it against interference from other connections.

Bluetooth is packet based and hops to a new frequency after each packet is received. This reduces interference while enhancing security. The data rates (with headers) are at least 1 Mbps, whereas full-duplex transmission (in both directions simultaneously) is achieved by using time-division multiplexing.

Bluetooth operates at 2.4 GHz (like 802.11b), which is the unlicensed portion of the ISM (industrial scientific medical) band. The frequency band is subdivided into 79 hops, 1 MHz apart. The 2.4-GHz band starts with 2.402 and ends with 2.470 (with narrower applications in foreign countries). The spread spectrum allows the data transmission to hop from one channel to the next in a pseudo-random manner. The idea is that by jumping randomly from one channel to another, it is difficult to eavesdrop on the transmission; this adds a much stronger layer of security. This means you may have as many as 1600 hops per second, and since the normal frequency range is 10 cm to 10 m this can be extended to as much as 100 meters when you increase the transmission power.

Bluetooth Connections

Bluetooth connections are established through the following means:

1. **Standby**—Any device not connected through a PAN is initially in standby mode. In this mode, devices monitor for messages every 1.28 seconds over 32 distinct hop frequencies.
2. **Page/inquiry**—When one of your devices needs to form a connection with another device, it transmits a page message. If it knows the address, then the inquiry is received along with a page message. The master unit transmits 16 identical page messages throughout 16 hop frequencies to the slave unit. If there is no response, the master unit retransmits on the other 16 hop frequencies. The inquiry method needs an extra response from the slave unit, due to the fact that the master unit does not know the specific MAC address.
3. **Active**—Data transmission takes place.
4. **Hold**—This occurs when either the master or slave must go into a "hold mode," when it doesn't transmit any data in an effort to conserve power. Normally, there is a constant data exchange so there can be hold mode in the connection of several PANs.
5. **Sniff**—The sniff mode works only in slave units and is used mostly for power conservation purposes. However, this mode is not as restrictive as hold mode. When functioning in this mode, the slave does not take an active role within the PAN. It does, however, listen at a reduced level. Note that this is a programmable setting and can be adjusted according to your needs.

6. **Park**—Park mode represents a significantly reduced level of activity below that for hold mode. During this time the slave is synchronized to the PAN, meaning it is neither required for full reactivation, nor is it a division of the traffic. In park mode, these units do not have MAC addresses; however they only listen to the traffic in order to keep their synchronization with the master unit and to check for broadcast messages.

Data transmission When dealing with data that can be sent either asynchronously or synchronously, however, you can use the synchronous connection oriented (SCO) mechanism mostly for voice communications, whereas an asynchronous connectionless (ACL) mechanism is meant mostly for data communications.

When working with a PAN, each pair between the master and the slave can utilize a different mode of transmission that can be changed on the fly. For example, time-division duplex (TDD) is employed by both ACL and SCO. Each of these protocols offers support for 16 types or packet flavors. Four of these packets represent the same thing in each type due to the need for an uninterrupted data transmission. In contrast, SCO packets are sent at periodic times, so that they are transmitted in groups without permitting any interruptions from other transmissions. Furthermore, SCO packets are sent without polling from the sending unit, while ACL links offer support for both asymmetric and symmetric transmission modes. The master unit actually controls the bandwidth in an effort to define how much of each slave unit can be used. This is possible due to the fact that slave units are not able to transmit data until it has been polled by the master unit. The master unit, however, can unit the ACL link to send broadcast messages to the slave units.

Error correction To ensure adequate transmission, there are three error connection methods used to ensure data is accurate:

1. One-third rate forward error correction (FEC) code
2. Two-thirds rate forward error correction (FEC) code
3. Automatic repeat request (ARQ)

The FEC mechanisms were created to reduce the number of repeat transmissions. Throughput is curtailed by slower transmissions when numerous repeats are requested. In a transmission environment not prone to errors (short-range WLAN) this method is not usually employed, but still has packet headers.

ARQ mechanism must receive the header error code (HEC) as well as the cyclic redundancy checks (CRC) so that when an acknowledgment is sent, everything proceeds normally; however when it is not sent then the data packet is transmitted over again.

Enforcing Security

Security is enforced through three primary mechanisms as shown in Figure 2.4:

1. Authentication
2. Encryption
3. Pseudorandom frequency band hops

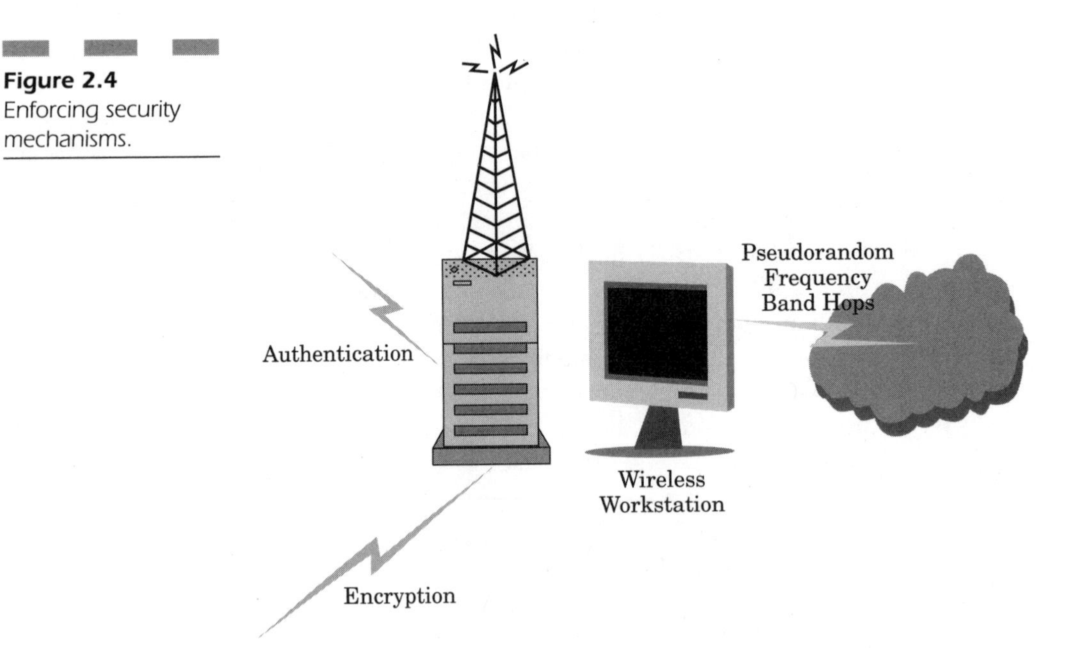

Figure 2.4 Enforcing security mechanisms.

The last mechanism, frequency band hops, causes the greatest problems for any hacker attempting to eavesdrop on your wireless communications. Authentication permits the user to control how device connectivity is designated for each individual wireless user.

In order to prevent anyone who can eavesdrop on your wireless communications from understanding or using your information for any unlawful purpose, encryption techniques use a secret key to scramble your information so it is unusable to anyone who doesn't know the key. Secret key length is normally 1 bit, 40 bits, or 64 bits.

The higher the bit level, the more security you have for your applications. Aside from using the highest level of encryption possible, it is just as important to have distinct types of transfer protocols monitored by wireless intrusion detection software. This all provides you with a superior level of protection regardless of platform, computer, or networking medium used.

Link Me Up!

Bluetooth systems are composed of both a radio chip and a controller that uses what is called link manager (LM) software to control link setup, authentication, and link configuration (Figure 2.5).

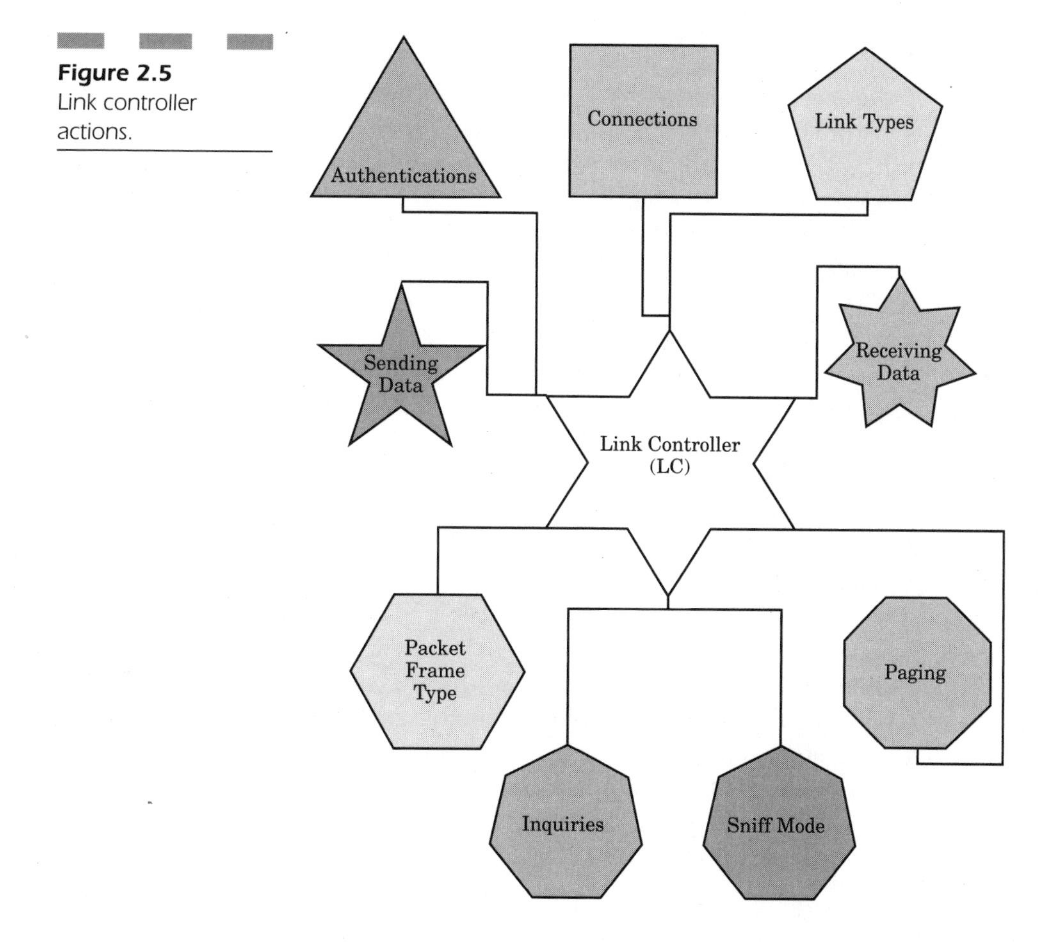

Figure 2.5 Link controller actions.

The hardware for the link manager is the link controller (LC) that executes the following actions:

1. Sending data
2. Receiving data
3. Authentication
4. Establishing connections
5. Establishing link types
6. Determining the packet frame type
7. Paging
8. Inquiries
9. Setting a device in sniff or hold mode

Conclusion: The Future of the WLAN

What will the future hold for wireless devices? At the moment, 802.11b is the more popular wireless technology. Bluetooth hasn't really caught on as much as 802.11, but it is the most competitive wireless alternative protocol to offer a really viable technological alternative for a mobile user's connectivity needs.

Since there are both pros and cons to 802.11 and Bluetooth, the two technologies may complement each other's capabilities as they proliferate. The main concern is speed. While Bluetooth is still basically maxed out at 2 Mbps, 802.11b can achieve 11 Mbps, and with 802.11a coming down in price and increasing in practicality (with compatibility to 802.11b devices) it may well be the winner in the next few years.

When we compare technologies, we can't make a direct comparison per se between wireless devices based on speed. Bluetooth is more advanced in the area of mobile phone devices and some PDAs because it was designed more as a universal standard that can be used to eliminate our wired world. 802.11 is more popular and provides users with needed speed. This means that mobile devices will more than likely stick with Bluetooth applications, while wireless workstations will stay with 802.11.

Support for 802.11 is more evident as Microsoft Windows XP, Mac OS X, and the new operating system called Lindows OS (Linux-based operating system that can run some Windows applications) all offer integrated support for 802.11 right out of the box without any configuration necessary. This is the best indicator that 802.11 is here to stay and will have the most impact on WLANs for the near future and beyond.

CHAPTER 3

Wireless LAN Security Factors

The main difficulty in establishing a wireless network is being able to support effective security so that users can access your network without fear of leaking mission-critical data through the airwaves in or near the perimeter of your office building.

Security of your WLAN remains an area of great debate and concern for the foreseeable future. This chapter examines the issues critical to WLAN users with respect to the following factors, as shown in Figure 3.1.

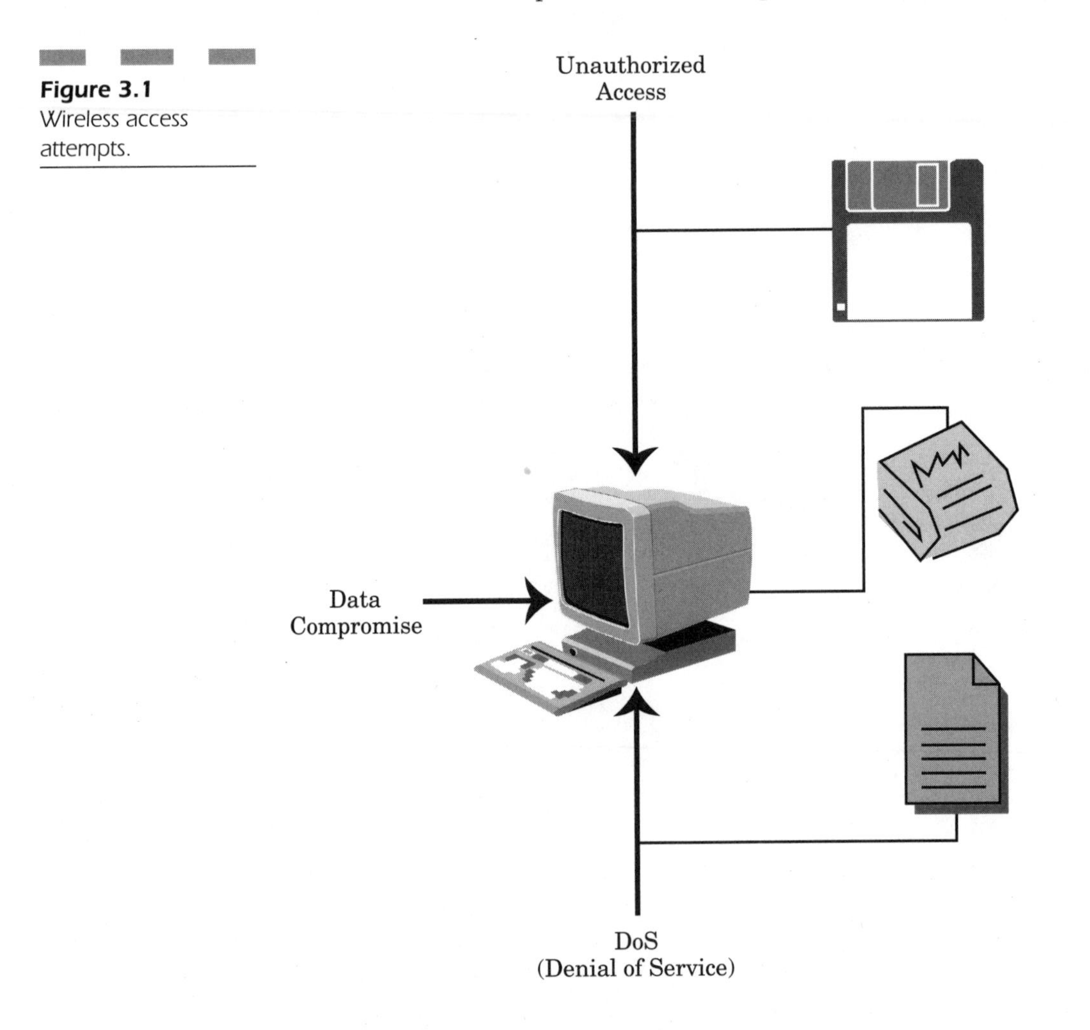

Figure 3.1 Wireless access attempts.

1. *Data compromise* is any form of disclosure to unintended parties of information. Data compromise can be inappropriate access to payroll records by company employees, or industrial espionage whereby marketing plans are disclosed to a competitor.

2. *Unauthorized access* is any means by which an unauthorized party is allowed access to network resources or facilities. Unauthorized access can lead to compromise if access is gained to a server with unencrypted information or to destruction, since critical files, although encrypted on the server, may be destroyed.
3. *Denial of service (DoS)* is an operation designed to block or disrupt the normal activities of a network or facility. This can take the form of false requests for login to a server, whereby the server is too distracted to accommodate proper login requests.

Enabling Encryption Security

The problem with most wireless LANs is that security is often considered optional and is turned off by default on every system. The entire premise of a wireless network is a wonderful convenience; however it has no security out of the box. It becomes your responsibility to determine how best to enable security so that people don't attempt to access your network without your knowledge.

Why don't most people enable security by choice? This is an important question that has a good answer. An 802.11b network, for example, with the best possible range and signal, has a maximum throughput of 11 Mpbs. While that speed may have been considered "as good as it gets" five to ten years ago, today people are finding wired 100 Mpbs LANs too congested for transferring files and other large objects over the network.

When you enable security on a wireless device, there is a certain degree of overhead that reduces the overall speed of your connection because it is effectively encrypting your network traffic on one end and decrypting it on another end. While the computer processes this information quite quickly, it cuts into your overall speed.

If you decide to enable a much stronger level of encryption in the 128-bit range, then you will have to deal with an even greater consumption of bandwidth involved when encrypting and decrypting your traffic. A greater portion of the radio frequency spectrum transmission is consumed with encrypted packets and this reduces your speed accordingly.

The 802.11b standard enables security through both authentication and encryption. Authentication is either a shared key or an open system. When the network router receives information, it may permit a request to be authenticated on that one station or on all the stations on its list. With a shared key, only those stations that have that same encrypted key can permit authenticated users to access that portion of the wireless network.

WEP Encryption

Wired equivalent privacy (WEP) encryption is the ability of 802.11 to create security that is analogous to that of wired networks. WEP uses the RC4 algorithm to encrypt wireless transmissions. However, WEP encryption does not cover end-to-end transmissions. It protects only the data packet information, but not the actual physical layer that contains the header. This means that other wireless stations on the network can receive the control data necessary to manage the network. The general idea is that other stations won't be able to decrypt the data segments of the packet.

Encrypting 802.11b?

The 802.11b specification is the means by which most wireless networks function. They work in the 2.4- to 2.48-GHz band as both ad hoc and extended service set networks. 802.11b, as opposed to 802.11, does not use FHSS as a mode of transmitting data. Instead, it establishes DSSS as the standard by which it transmits data because that is much better in handling weak signals. DSSS allows data to be much more easily distinguished from the interference in the background without the need to be retransmitted again. Due to the strengths of DSSS over FHSS, 802.11b is able to reach general speeds as high as 11 Mbps at close range with a slightly lower rate of speed of 5.5 Mbps at 25–50 feet away from the transmission source indoors.

802.11b has evolved to the point where the majority of hardware can support 64- to 128-bit encryption schemes, whereas 802.11 only supported 40- and 64-bit encryption.

Network Interface Cards

Wireless network interface cards (NICs) now offer the stronger 128-bit encryption schemes and have their own unique media access control (MAC) addresses that identify that card on your network. They also possess their own public and private key pairs to maintain a straightforward method of encryption to the WLAN. These unique identifiers allow you to use these MAC addresses as a means of access control to allow only specified NICs onto your network. You can program your access points (wireless routers) to look for this unique hardware NIC identifier and permit only that address to access your network resources. This entire exchange is completely transparent to the user, doesn't consume any extra network bandwidth, and maintains a higher level of security. Should someone attempt to log onto your network with an unauthorized network card from the parking lot in front of your business offices, the access point would automatically determine that the MAC address of the hacker's wireless NIC card is not on the authorized user list and deny access.

You can also program some wireless routers/access points to maintain a log of all the MAC address combinations they see and then reject any addresses they don't recognize. This method allows you to prevent a hacker from attempting to break into your network by trying to spoof the MAC address of his NIC card to emulate an address of a card that *is* authorized to access your network.

Cross-Platform Hacking

The sad fact is that no matter whether you are using a Windows, Macintosh, Linux, or Windows CE PDA, you are as vulnerable to the same types of hacker attacks as wired networks, only you have to worry about these attacks coming from outside your building.

It is not uncommon for hackers to try and access your WLAN, but did you know they don't have to be located right outside your building? In fact, many workers are finding that with the proper antenna array you can access an 802.11b network from as far as a few miles away from your offices. Parabolic dishes (like the ones used for satellite television) can be aimed from one point to another. Many mobile workers who live near their offices have one of these 18-inch dishes on the roof pointed in a clear line of sight to another parabolic dish on the roof of someone's home. It is possible someone can be *that* far away and still have access to your system.

Hacking is not just affected by distance; it can be done from one platform to another. One very important fact about 802.11b, and one which people are unaware of, is that the Macintosh computers in both the old

Mac OS 8.x–9.x and the new Mac OS X use an "Airport" card. The Airport card for the Macintosh *is* an 802.11b card! You may be thinking that the Mac is a different platform and can't access the resources on my Windows network, so I'm safe. *Wrong!* There are a few programs for the Macintosh that enable that computer to access any Windows network share through the native Macintosh interface. One program that works very well is DAVE, from Thursby Software. This program works in the older Mac OS 8 and 9 as well as natively integrating itself into Mac OS X. It makes access to a Windows share on your wireless network very easy. In fact, it can even make a Mac computer appear to be a Windows NT type of workstation on your Windows network. That means someone could theoretically be inside your company dragging and dropping files onto what appears to be another Windows computer, but could very easily in fact be a Macintosh computer running DAVE. The Mac integrates even more easily into a WLAN than a PC does!

You might not think Linux-based machines running common distributions (Red Hat, SuSE, and others) supported 802.11b, but they do! In fact, both Red Hat and SuSE have built-in native support for at least five common wireless NIC cards on the market. Linux is a great operating system because not only can you access wireless network shares on 802.11b, but there are a wealth of very easy to use and comprehensive hacking tools on the Internet that allow the Linux machine to monitor your wireless network, probe its traffic, and create an entire picture of everything you have running. Accessing anything on your network from Linux is almost as easy as using a Macintosh; but with more hacking utilities, Linux becomes a much more dangerous adversary of your wireless security.

The most difficult device to detect on your network is the wireless PDA. Both Palm and Windows CE devices can now natively support 802.11b. They can access e-mail and Web shares, and there are a number of utilities that allow them to emulate an entire Linux or DOS operating system to gain pocket functionality to hack into your systems. For the Windows CE environment, a program called "PocketDOS" can not only emulate a full DOS system on your PocketPC or Windows CE devices, but can also emulate Linux in a convenient portable package. With the proper knowledge, these devices can be totally concealed, yet have complete wireless access to every system on your wireless network without anyone's even knowing about it!

Just as wireless LANs permeate building materials and distance, they can also migrate from one platform to another. The truth is that no matter what you do, you can never calculate all the devices that will have access to your WLAN. In fact, the new generation of cellular phones have a PDA built right into them, and soon will include either BlueTooth or 802.11b built directly into these devices. The next time you see someone take out their cellular phone to make a call, they could very well be connecting into your wireless network. Most locations in airports across the country even have wireless LANs set up to allow people to have mobile Internet access at any time while waiting for their flights. The only problem with this convenience is that your traffic is almost certainly not encrypted (because these types of networks were designed to be public without any encryption methods whatsoever). This means that it is a very simple matter for someone to eavesdrop on your network connection and see any or all of the corporate data you are sending to your home network.

Eavesdropping

Hackers can easily eavesdrop on your network traffic by monitoring the radio waves transmitted by your access point or wireless router. While this type of attack is considered passive in nature, it is simple to accomplish. All the hacker really needs is a radio receiver with a high-gain antenna that can intercept transmissions of network traffic. This type of attack can take place without the knowledge of either the network administrator or the user. For all intents and purposes, there is really no straightforward defense against this type of attack except to limit the range of transmission from your wireless access point.

The inherent security that 802.11 offers is by design. Although there are a number of pieces of equipment on the market designed to intercept WLAN network traffic, such interception is not easy. 802.11 uses digital spread spectrum on the 2.4-GHz frequency, meaning its transmission is spread throughout the band, making it that much harder to pinpoint the signal and eavesdrop. Moreover, if you enable encryption on your wireless access point, you are in a better position to resist anyone's eavesdropping on your signal because even if a hacker does listen to your signal, he would have to decode the transmission before making any usable sense out of it. However, since encryption doesn't pose much of an obstacle to a hacker, eavesdropping should be considered a deadly threat to the safety of any mission-critical information transmitted over the network.

Breaking In!

An active type of attack is when a wireless user actually breaks into your network disguised as an authorized user. Even if you have taken the precautions of encrypting your network traffic and blocking out any unauthorized wireless NIC cards, a hacker could potentially steal an authorized wireless NIC card or possibly bribe someone with after-hours access to add the MAC address of an unauthorized NIC card into the authorized list of users that the access point will accept.

Once the hacker gains access to your internal systems, he can corrupt, steal, erase, or destroy confidential data pretty much anywhere in your entire network. The hacker could potentially have access to your systems for a long time if left unchecked, and could be stealing important presentations, market information, pricing data, or research and development information directly from your network for an extended period of time. This type of attack is not uncommon.

The only way to combat a hacker is to have someone attuned to your network bandwidth with extensive knowledge of all the users authorized to access your wireless network. You must be very careful about who has access to your information and during what hours this access occurs. There are several methods of detection; the most common is to monitor and log all your WLAN activity for access during off hours. During business hours, you can check to see if there is an unusual amount of network congestion, caused by a hacker consuming all available network bandwidth while copying important data files from your server directly. Most companies keep a log of at least 28 days, since only logs of extended periods of time show any intrusion detection attempts to access your system from an off-site location.

Detecting unauthorized attempts to access your WLAN is often complicated by the fact that this medium (by design) has a high bit error rate (BER) which often makes it appear that intrusion attempts and unsuccessful access attempts are one and the same. When an access attempt is not successful, this action is often seen as simply an unsuccessful logon attempt. This makes it more difficult to track down intrusions on WLAN than on wired LANs.

Counterfeiting

In counterfeiting, a hacker sets up an unauthorized access point to make other wireless stations access it instead of the authorized network. When a wireless user moves from one location to another, the NIC card often latches onto the strongest cell in its area of reception, much as a cellular phone moves from cell to cell, switching to the one with the most power and greatest signal strength. The counterfeit access point can attract a wireless station into the false network in order to copy its encryption key used to log on to the real network access point. In addition, the user would normally send his password to log onto the network; the counterfeit access point would capture that too. The counterfeit systems may actually be much farther away, but it is a simple matter to reconfigure most access points to increase their output power beyond the legal limit to attract a greater number of wireless stations anywhere in their vicinity.

A counterfeiting attack is difficult and requires a greater level of knowledge about the access point and protocols of the wireless corporate network being imitated. Without detailed knowledge about the internal network, wireless users would immediately see something is wrong, making this type of attack easy to detect. It is hard to track down these types of attacks because all that is really needed to pull this off is a receiver and antenna compatible with the targeted wireless stations. It is difficult to detect this attack (when it is taking place) because unsuccessful logons are extremely common in the WLAN environment.

The only way to truly protect yourself against a counterfeiting attack is to implement a strong and efficient means of authentication that requires wireless stations to authenticate themselves to the access point while leaking neither the shared cryptographic key nor the passwords to access network resources.

Wireless DoS Attack

If all else fails and a hacker simply wants to disrupt your wireless network, he can create a wireless denial of service (DoS) attack that renders your entire wireless network unusable. This is accomplished by creating a transmitter powerful enough to flood the 2.4-GHz band (the frequency spectrum that 802.11 uses to make WLAN connections) with interference. With sufficient power, this type of attack can render any wireless network traffic null.

These types of attacks can take place from a car parked near your office building, the rooftop of a neighboring building, or a sufficiently powerful line-of-sight transmission from as far as a few miles away. The problem is that if your offices use wireless networks over your entire corporate premises, you could lose work and connectivity because someone is trying to destroy your ability to do business effectively.

Points of Vulnerability

Beyond wireless DoS attacks, there are several points of vulnerability within your WLAN that can be disrupted or destroyed by knowledgeable hackers who are trying to corrupt your wireless infrastructure. The most vulnerable points include those shown in Figure 3.2. These points are:

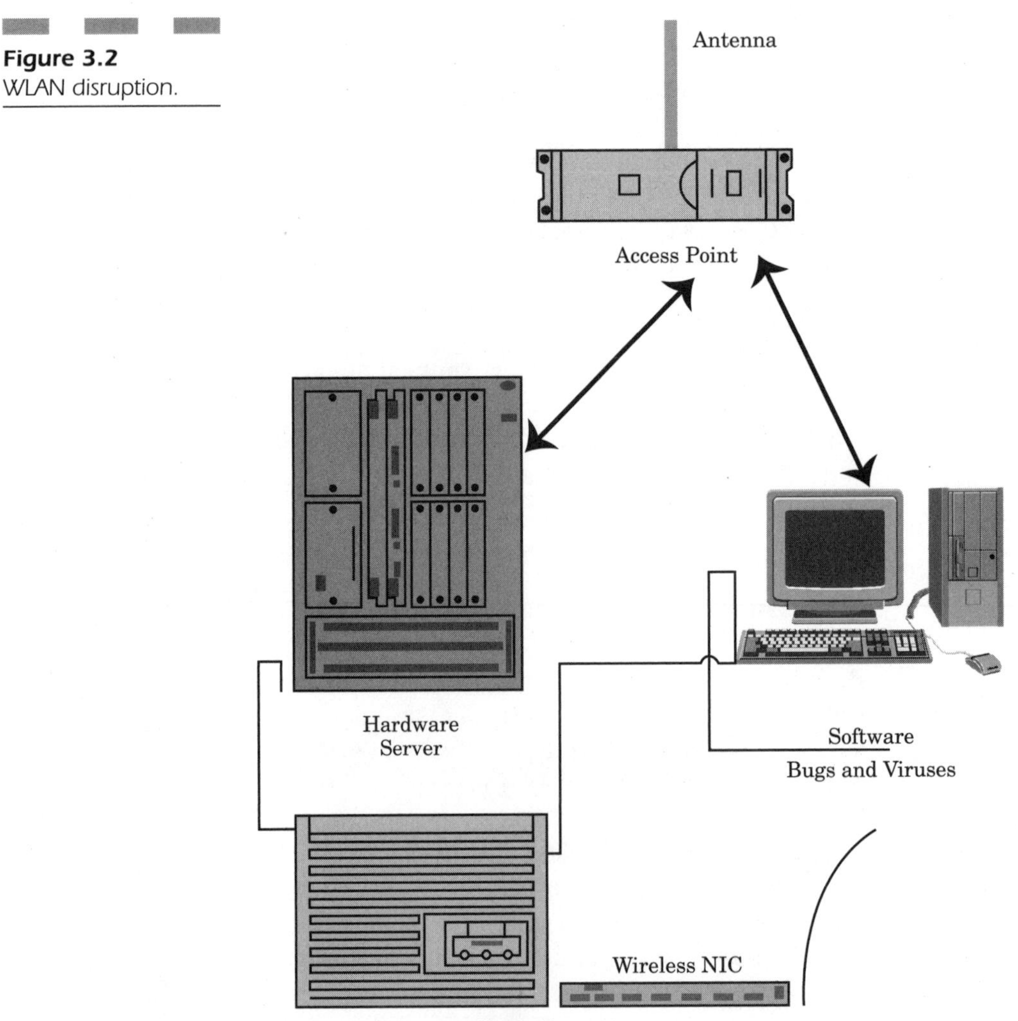

Figure 3.2
WLAN disruption.

- Access points
- Antennas
- Wireless NICs
- Cable connectors
- Hardware servers
- Software bugs and viruses

It is simple to infect software from virtually anywhere. However, what is not commonly known is that hackers can send firmware upgrade attacks to your wireless router and access point. The firmware in these devices is a software file that updates your device to take advantage of new features and functionality. If a hacker gains access to this device, he can rewrite a valid firmware file or simply corrupt it and fail to load the firmware correctly onto the access point, thus rendering the device completely unusable.

Servers and software can be infected by any number of viruses, but most newly made viruses can look for adapter connectivity and wireless network adapters to corrupt the means by which the server communicates with these devices on your network, as shown in Figure 3.3.

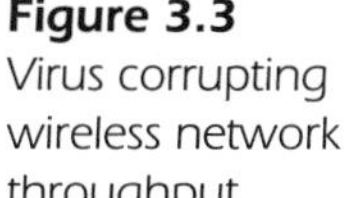

Figure 3.3 Virus corrupting wireless network throughput.

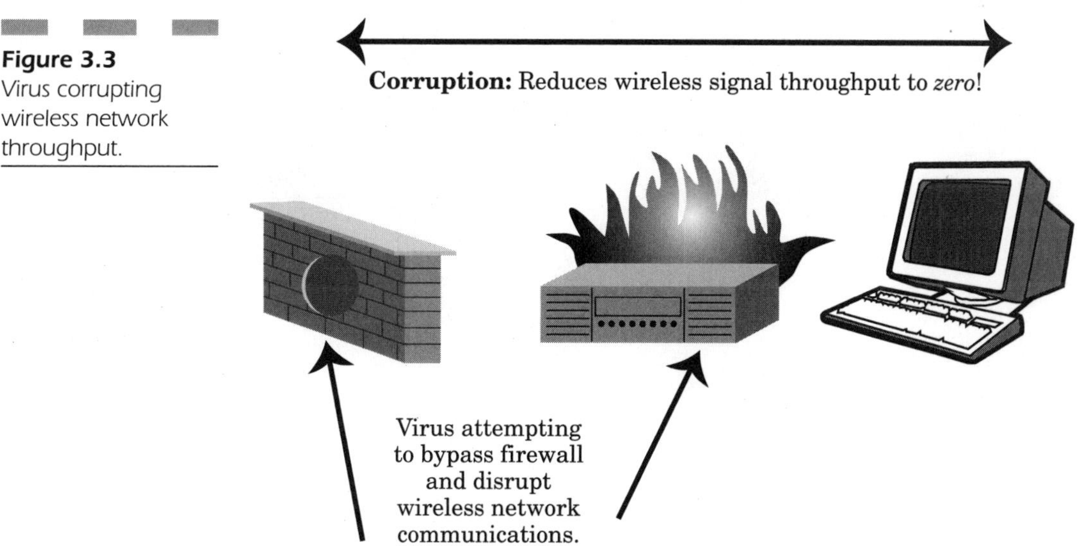

The goal of these types of attacks is either of the following:

1. Complete shutdown of your wireless networking devices
2. Corrupting your signal to reduce throughput to zero

Shutting down your wireless networking devices cuts off your entire network, but your company can easily purchase a new component. However, if the hacker corrupts your WLAN so that your throughput is greatly reduced, it is far more difficult to determine whether the problem is hardware or software related. In the meantime, users have such slow traffic on your network that your WLAN becomes virtually unusable.

Shutting down your network can involve something as simple as having the hacker gain access to your premises or having someone access your premises on his behalf (such as a janitor or cleaning crew who was paid money to sabotage your network systems). See Figure 3.4. Your network can be sabotaged by:

- Disrupting the connections between access points
- Cutting the connection from the wired LAN to the WLAN
- Isolating various access points so that they cannot communicate from one cell to another, thereby cutting your overall reception
- Cutting power to one or several access points

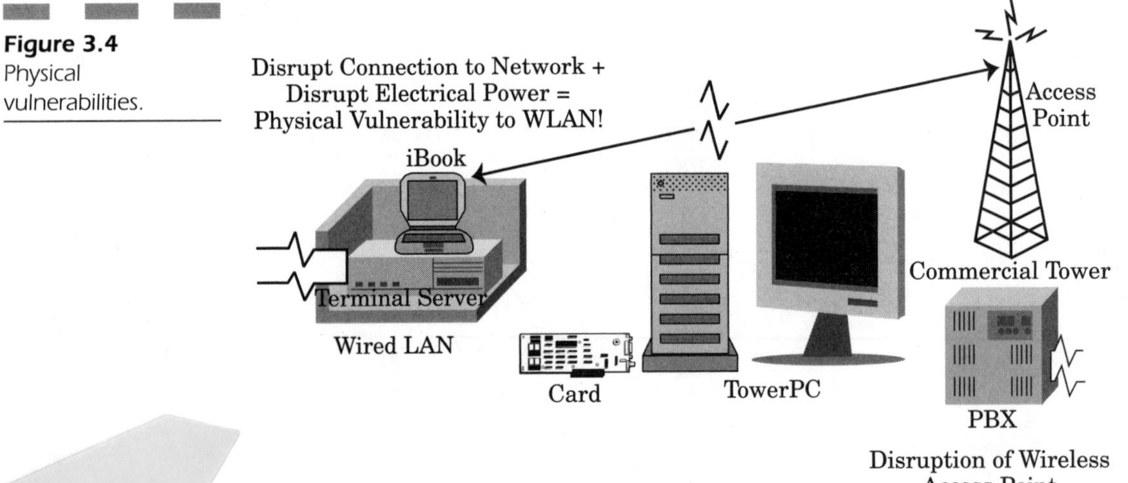

Figure 3.4 Physical vulnerabilities.

In addition, it is common for a hacker to use a registered wireless NIC transmitter to cause interference and disrupt network traffic. Furthermore, disrupting any connection to the server, wired LAN, or network resources can destroy the validity of your network in a number of ways causing damage, lost time, and lost work.

Your Best Defense Against an Attack

802.11 uses spread-spectrum technology, which sounds almost like background noise to the average person. However, someone skilled in eavesdropping techniques can determine the transmission parameters of the 802.11 signal in order to decode the spreading code and put it into usable form.

One form of protection is to shield your facility by limiting the range of your wireless equipment to those inside your corporate facilities only. See Figure 3.5.

For example, frequency-hopping spread spectrum (FHSS) hops over 75 different frequencies with respect to a somewhat random code sequence that both the transmitter and receiver lock onto. There are 22 distinct hopping patterns, selected by the transmitter using a designated type of code. The receiver can detect a hop pattern and then synchronize to the transmitter. The idea is to keep the pattern changing by resetting the devices at specified intervals. This is one form of defense to protect your FHSS pattern from being detected and used to listen in on your network traffic.

In direct-sequence spread spectrum (DSSS), each data bit is segmented into the signal in chips that are then migrated into a waveform transmitted over several different frequencies. The receiver then blends the chips to decode the original data signal. 802.11b uses 64 eight-bit code words to segment the signal. When trying to listen in on that signal, the hacker sees the DSSS signal as background wideband noise. Your defense is to try to use several DSSS signals to make it appear that you have overlapping 802.11b devices. While this may not prevent eavesdropping, it makes it difficult to pick out one access point among many.

It is sometimes best to use a combination of the these two types of systems to confuse any would-be attacker; this results in a better method of defending your system against hackers interested in eavesdropping on your systems. When dealing with FHSS, the hacker needs to know the hopping patterns used in your wireless transmissions. When dealing with systems running DSSS, the hacker needs to know the chipping code or code words present in either 802.11 or 802.11b. In addition, regardless of which method you deploy, the hacker must know the frequency band and modulation to decode the transmitted data signal correctly.

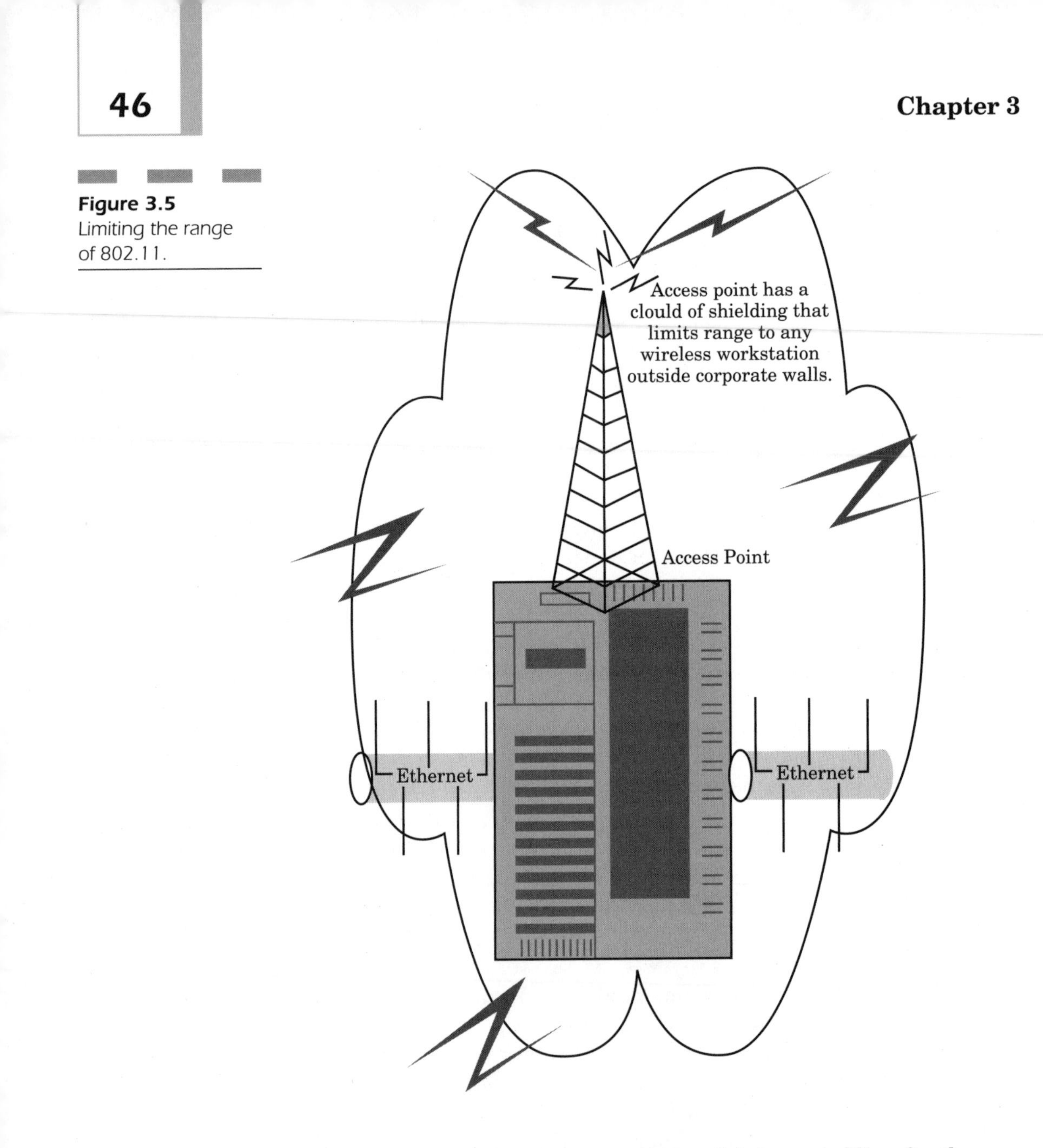

Figure 3.5 Limiting the range of 802.11.

Since radio transmissions use a type of data scrambling for the purpose of better timing and decoding of radio signals, the hacker must know the specific pattern that he needs to decode information intercepted from your WLAN. Another benefit in defending you is that neither FHSS nor DSSS is interoperable; even though these two different types of systems are using the same type of wireless transmission, they are not able to communicate if they are using different frequency bands. DSSS is not able to talk to another system using DSSS if they are functioning on two different frequencies. In addition, the hacker cannot use any given spread-spectrum type of attack to intercept radio transmission by any other mode of transmission. The hacker is also not able to intercept radio transmissions without knowing the exact frequency used, regardless of whether he or she owns a compatible 802.11 receiving device.

The main factor in keeping 802.11 secure from hackers is to make certain that your hopping pattern or chipping code is not known to the hacker. If the hacker does gain knowledge of these parameters (which are published in the 802.11 standard) he could devise a method to determine your modulation. This information can provide the hacker with the ability to create a receiver to intercept and read the signals from your network.

There are numerous benefits in your spread-spectrum technology that make it very difficult for the majority of interested hackers, so 802.11 is a reasonably secure platform for your WLAN.

The entire concept of spread-spectrum technology is to reduce the amount of interference from other radio devices by spreading radio signals over a huge range of frequencies. However, it is still possible for a hacker to jam your signals. Your defense against this type of attack is to insulate the exterior of your building so that radio signals from outside the walls of your corporate WLAN have great difficulty in penetrating or disrupting your network. This defense works two ways; insulating your walls with shielding materials not only blocks out jamming devices, but also serves to isolate your WLAN and make it much, much more difficult to eavesdrop or log onto your network from any great distance beyond your parking lot.

One of the more interesting defenses of your WLAN is to avoid using radio waves in favor of using infrared types of transmissions. You can use the same type of wireless connectivity, but you need to be in range using line of sight to the infrared transmitter. There are numerous limitations to these types of transmissions, but it is valid to point out that with a good bit of strategy and placement you can effectively make it exceedingly difficult for someone to compromise your WLAN.

Conclusion: Keeping Your WLAN Secure

Wireless LANs pose a serious security threat for those companies that believe that the technology alone (out of the box) will ensure security for wireless corporate users. In fact, this is a relatively insecure medium that has a great number of potential holes that not only can leak your mission-critical network traffic, but potentially allow someone to gain unauthorized access to your network from outside your building.

In this chapter, we have seen that 802.11b is a shared protocol used by Windows, Macintosh, Linux, and numerous wireless PDA devices. With so many platforms existing in the same wireless protocol, think about how many attack patterns are possible in compromising the integrity of your WLAN.

Your goal is to remain as vigilant as possible in ensuring the security of your wireless network. Make certain you turn encryption on for all your wireless stations and access points. Make certain to use the highest strength of encryption possible in order to make it as difficult as you can for a hacker to gain access to your network or eavesdrop on your network traffic. Don't be fooled into thinking that the lowest level of encryption (40- or 64-bits) is sufficiently high to stop a hacker; it won't. If someone is *really* interested in accessing your wireless network resources, given a small amount of dedicated time, your network (even at 128-bit) encryption *will be compromised*! However, if you make certain to change your key parameters at regular intervals and make certain you are aware of the different encryption keys you use, then you are in a better position to keep your WLAN safe.

Finally, know that each wireless network interface card has its own unique machine or MAC address. You should always configure your wireless router or access point to accept only connections from NIC cards that you have preauthorized for the network. This ensures that a hacker will have greater difficulty in accessing your network using a "parking lot attack" to set his wireless NIC card into promiscuous mode to log onto your network.

It is important to note that while no wireless security solution is 100 percent effective, you can take these very simple preventive steps to ensure that your WLAN is as secure as possible. When a hacker tries to intercept your network data or compromise your system, the more difficult you make his job, the more likely it is that you will have time to detect the attempted incursion into your system and prevent it. Protection is your best defense when it comes to 802.11!

CHAPTER 4

Issues in Wireless Security

This chapter presents an assessment of wireless security with focus on the effective response to the three primary issues noted below:

- Is the data adequately protected from compromise during transmission?
- Is access to the transmission and other information on the network controlled?
- Is there adequate protection from the range of DoS attacks?

The specific features of the RF transmission involved are also an issue since emanations are accessible to unintended recipients:

- What frequencies are available?
- How much transmitter power is required to ensure successful receipt?

We examine how security is applied in the wireless LAN and determine how these issues affect your environment. The idea is to see what pertains to your setup so that you can understand and effectively deal with these issues in your wireless security before they become a problem.

The State of Wireless LAN Security

In order to convince you that there are real issues to consider when implementing your WLAN, it is important to focus on the integrated security features present within 802.11b and their limitations.

802.11b offers features and functionality that provide you with greater security in your wireless environment, however these security services are enabled for the most part through the wired equivalent privacy (WEP) mechanism to protect you at the link level during wireless transmissions that take place between the client and the access point. Note that WEP is not able to offer end-to-end security, but it does attempt to secure the actual radio transmission by encrypting the data channel.

Securing Your WLAN

The most important issue when dealing with wireless security is to consider the fundamental security mechanisms in your wireless network. There are two primary means of adding security to your environment (Figure 4.1):

1. **Authentication**—This mechanism has the objective of using WEP to enable your security to be verified by determining the actual information that defines each wireless workstation. It is necessary to yield access control to the network by restricting wireless workstation access to those clients who can properly authenticate themselves to the server.
2. **Privacy**—WEP maintains an effective level of privacy when dealing with security for the data communication channels in your wireless network. It attempts to stop information from being "hacked" by attackers trying to eavesdrop on your data transmissions. The objective is to make certain that messages are not altered while moving from the wireless workstation to the access point or server. Essentially, this is the means that enables you to trust your information so that you can be reasonably certain your information is secure and reliable.

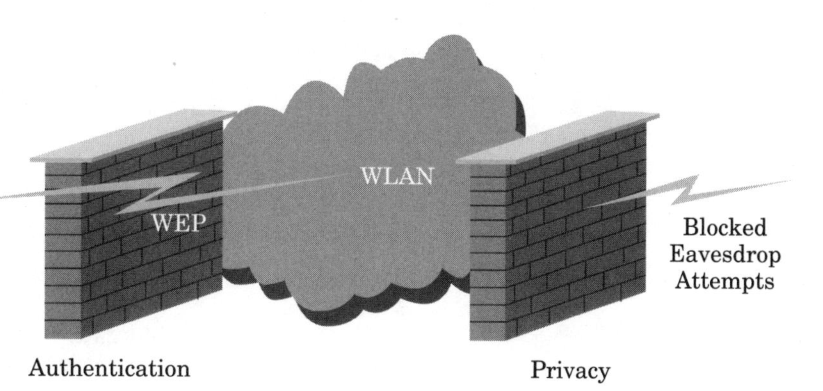

Figure 4.1 Securing your WLAN.

Authenticating Data

When a wireless user attempts to acquire access to your wired network infrastructure, there are two ways in which access can be obtained:

1. **Open system**—Any user in range of the access point can roam onto the system (as long as the router is not set up to filter out the unique MAC address of wireless workstations that are not supposed to have access).
2. **Encrypted system**—All data is scrambled and access barriers are put into place so that a hacker cannot eavesdrop on your data (Figure 4.2).

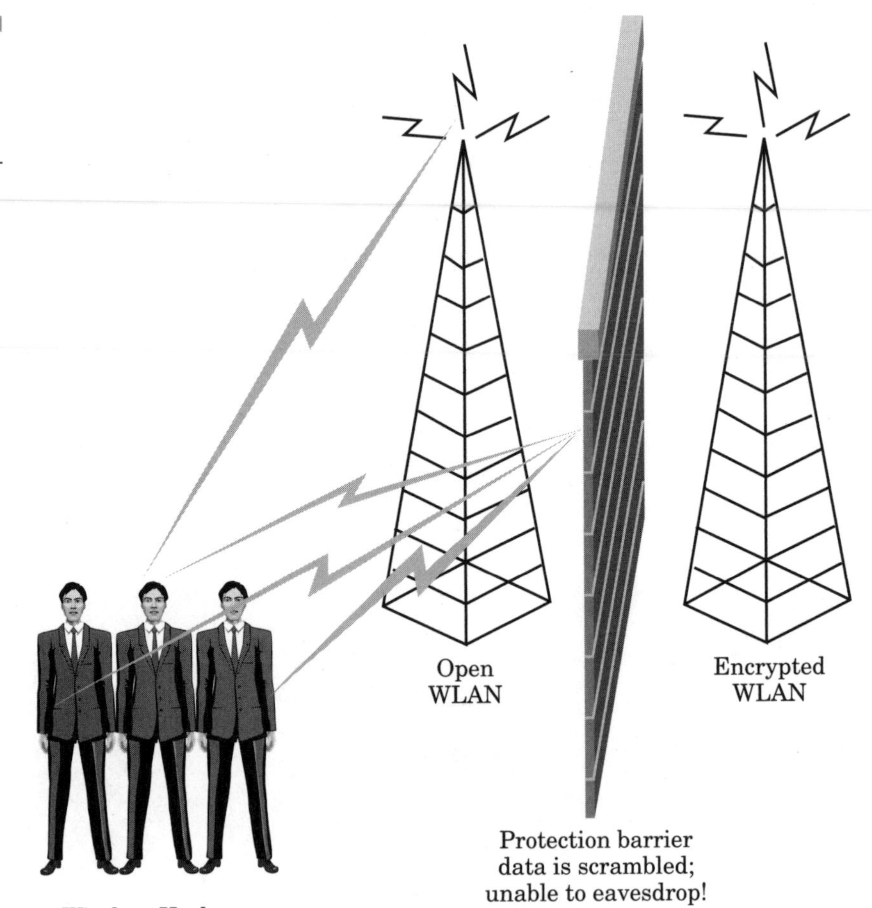

Figure 4.2 Protecting barrier safeguards network data.

In an open system without encryption, a wireless workstation can join your WLAN by using identity types of verification methods. The actual access request in an open environment occurs when the wireless server replies with the service set identifier (SSID) for the WLAN. This means there isn't any actual authentication taking place; the wireless workstation simply roams onto the network.

In contrast, you can see the differences spelled out between an open versus closed system:

	Open System	Closed System
Encryption	Nothing	RC4
Authentication	No SSID	SSID

Because of the unique SSID set for a company, many people believe that nobody could actually roam onto a network without knowing what unique identifier defined the network. In fact, it is possible for a wireless user to leave the SSID as "NULL" or blank; then when he is in range of the access point, the wireless workstation automatically finds and logs into the network. This means that basic systems of authentication are not sufficient to protect your network. This is why a combination of encryption and authentication is important in implementing your wireless security—but this still represents a small part of what needs to be done to provide a truly secure WLAN.

Client Authentication in a Closed System

In the previous section we saw that when a wireless workstation replies to the access point with a null or empty string in place of the actual SSID, it is automatically authenticated into the open system. However, when working in a closed authentication environment, the wireless workstation must reply with the exact SSID in order to log into the wireless network. The client is only granted access if it replies with the exact SSID string that identifies the client to the server.

Shared Key Authentication

The shared key authentication encryption mechanism uses the "challenge-response" mechanism. The idea is that each wireless client has an understanding of what is commonly referred to as a "shared secret."

The access point creates a random type of challenge that is transmitted to the wireless workstation. The wireless workstation then uses the encryption or WEP key it shares with the access point. The challenge is itself encrypted and then replies with the answer to the access point, which then deciphers that answer sent by the client. Based on the result, the client is granted access only if the deciphered answer is the same expected value as the random challenge.

RC4

Data is encrypted using the RC4 cipher. Note that the wireless workstation does not authenticate the access point, so that there is no verifiable means to make certain that the client is effectively talking to an authorized access point on the WLAN.

The problem is that it is possible for attacks to occur when hackers attempt to "spoof" authorized access points in order to "trick" wireless workstations or mobile users into inadvertently connecting to the hacker's access point, thus compromising the wireless network and stealing important information.

Ensuring Privacy

In dealing with security and privacy so much in my career, I once learned the mantra that "A security solution *without* ensuring privacy is *not a solution at all!*"

As we concentrate on the issues pertinent in wireless security, it is imperative to deal with the issue of privacy. The 802.11 standard can deal with privacy issues through using cryptographic mechanisms in its wireless connectivity.

The WEP mechanism ensures privacy through its use of the RC4 symmetric-key cipher algorithm to create a pseudorandom data sequence. WEP makes it possible for data to be protected from interception (or really understood) between transmission points along the wireless network (Figure 4.3). WEP is useful for all data in the WLAN, to protect and make your data channel private. The idea is to protect data when flowing through:

- Transmission control protocol/Internet protocol (TCP/IP)
- Internet packet exchange (IPX)
- Hyper text transfer protocol (HTTP)

WEP is designed to permit privacy by supporting cryptographic keys ranging in size from 40 to 104 bits. The idea is that by increasing the size of the key, you proportionally increase your level of security. For example, a secure setup includes a 104-bit WEP key using 128-bit RC4.

In practice, when you employ a key size in excess of 80 bits, it makes brute force hacker attacks very lengthy, time consuming, and generally unrealistic as a form of breaking into a network without being detected. In fact, with 80-bit keys, the number of possible keys is so great that even the most powerful computers produced today would not be powerful enough to break the code.

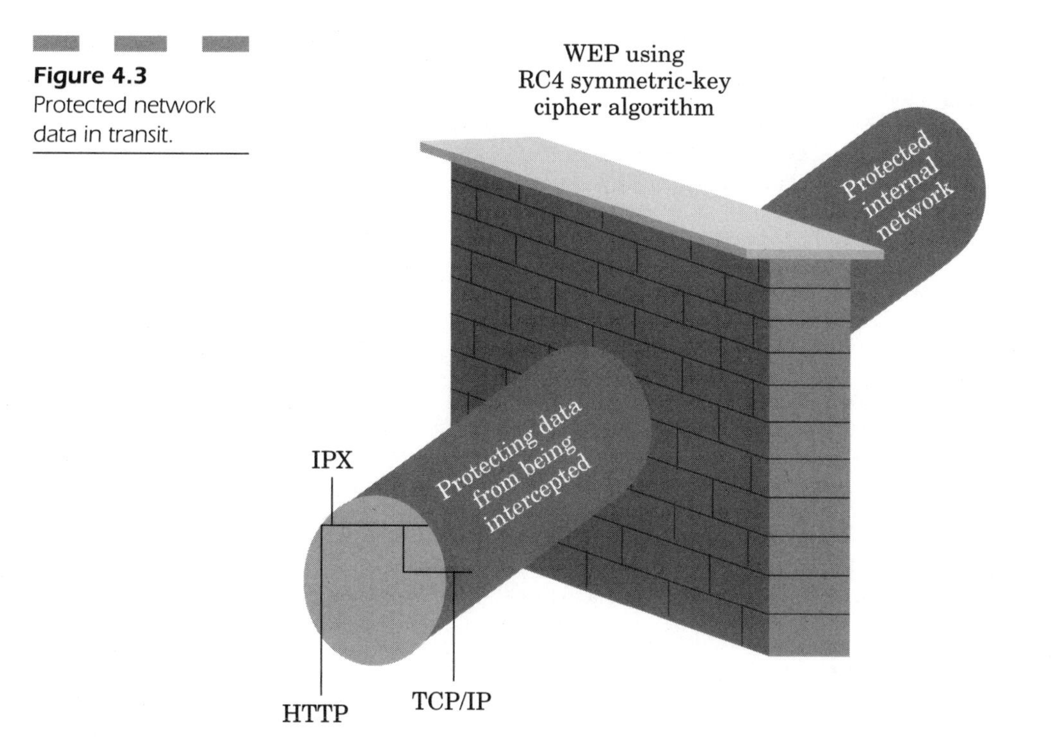

Figure 4.3
Protected network data in transit.

Unfortunately, in my experience, most companies don't use these keys for even the simplest form of protection on their network. Most WLAN implementations use only 40-bit keys. Most hacker attacks are successful on implementations that use 40-bit WEP keys; the majority of WLANs are at serious risk of being compromised.

Keeping Data Intact

One of the advantages of 802.11b is that it ensures that your data transmission remains intact as it follows the wireless path between the wireless workstation and the access point. The idea of this level of security is to reject any message transmission that may have been modified or intentionally altered during its path from point to point.

To maintain privacy, the 802.11 standard was designed specifically to reject any message altered in transit, either by accident or by design. To ensure that data privacy has been maintained, the cyclic redundancy check (CRC) technique is used as a form of encryption. This setup requires that each encrypted packet is "sealed" in a bubble using the

RC4 key encryption to scramble the transmission. Only when the packets are received are they decrypted; a CRC check is computed to ensure that it matches the CRC value before it was sent. Should the CRC value not match, then you have a receive error that defines an integrity violation and the packet is thrown away as corrupt.

Managing Keys

One of the problems with the 802.11 standard is that it has no good way of managing keys (Figure 4.4). The administrators who take care of your wireless network are responsible for several methods of managing keys with respect to:

- Creating keys
- Distributing keys among wireless users
- Archiving/storing keys so that they don't fall into the hands of a hacker
- Auditing who has what cryptographic keys
- Terminating keys that have become compromised

What happens if nobody takes care of these key management issues? Your wireless network is highly vulnerable to a hacker attack. These insecurities include:

- WEP keys are not unique and can be compromised
- Factory default passwords are prominently posted on hacker sites. This means that no matter which access point you are using, you are *vulnerable* if you have left your default administrative password unchanged since deploying your WLAN.
- Bad keys. Never make a key all zeros or all ones for the sake of convenience. Those types of keys are the first detected by a hacker looking to see how easy it will be to gain access to your wireless network.
- Factory defaults must always be changed as they are the easiest and simplest ways for a hacker to gain access.

The greatest difficulty is that the problem with managing keys grows in proportion with the size of your organization and the number of keys you will need to keep track of your wireless workforce.

Figure 4.4
Key management.

To indicate how extensive the task of managing keys actually is, consider that it is very difficult to scale your organization to change keys often enough to randomize them sufficiently to protect you against a hacker attack. In a large environment, you could be dealing with tens of thousands of keys.

In essence, vigilance and time are required, besides the fact that you must know how to protect your WLAN through the effective management of your encryption keys.

WLAN Vulnerabilities

There are a number of security vulnerabilities in 802.11 that have unfortunately been discovered by malicious hacker exploits. These vulnerabilities constitute passive types of attacks that are designed to decrypt traffic with respect to algorithms based on statistical analysis and active attacks designed to decipher network traffic. An active attack is basically accomplished by confusing the access point to give up to the attacker information it should not. This is the reason why default passwords and settings should always be changed as soon as you deploy your WLAN.

The most significant problem rests with WEP, which was itself designed to make a wireless network nearly as secure as the wired Ethernet. The biggest problems result from using the same WEP key over and over again. The more you use the same keys, the greater the chance an attacker will learn this piece of information so that he might ultimately use it against you for the purpose of accessing your WLAN. The vulnerability here rests in the fact that the same key is used for extended time periods, and nobody really thinks to change it. When you think of a WEP key, you should remember to change the key as often as you might change your logon password.

The *initialization vector* (IV) constitutes the 24-bit field transmitted in clear text as part of WEP. This 24-bit information initializes the RC4 algorithm key string. The IV is basically a short field used for encryption.

The IV is meant to protect your information, but a short IV ultimately gets repeated many times over the network when there is a great deal of traffic. The problem is that an attacker may easily use this information to intercept your wireless data channel, find your key stream, and then use this information to decipher the encrypted data on your WLAN.

Since the IV is actually an element from the RC4 encryption key, once the hacker has intercepted this bit of information and can intercept every packet key. Since the RC4 key is weak in and of itself, this could indicate the precursor of a significant attack. In fact, this attack could easily be run a script kiddie because once the secret key is recovered, it is possible to analyze only a small portion of the wireless network traffic and be able to have full access to the WLAN.

There isn't any protection for the actual composition of the encryption that WEP has to offer except that the MAC portion of the 802.11 standard uses the CRC element described earlier as a form of privacy protection.

Subtle Attacks

Another problem possible on your 802.11 WLAN is a WEP attack where a hacker initiates an active attack while simultaneously deciphering data channel packets by altering their information and CRC and then transmitting these altered bits of information back to the access point.

There is a great deal of risk associated with the creation of encryption protocols that do not possess a cryptographic privacy protection mechanism due to the communication necessary with several other protocol levels that can leak information about your encrypted data.

Common Security Pitfalls

Knowing the most common problems with WLAN security as it relates to the 802.11 standard can help you find and solve the problems with your implementation before they become vulnerabilities that hackers can exploit to your disadvantage.

Poor Security, Better than No Security at All!

The most common problem is that the security controls in your wireless equipment are turned off by default out of the box. Although these security features and functions are not all-encompassing to stop hackers, leaving them disabled just puts you at unjustified risk. Better that you should have minimal security measures as opposed to having no security enabled.

Short Keys

Most cipher keys are very short; most implementations use only 40-bit encryption keys, which can make the key stream repeat. There is no reason why you should not at least use larger key sizes when employing encryption techniques. To that end, a key size should be at least 80 bits long. When using longer keys, the likelihood of having them compromised by a hacker is far less. Hackers use "brute force" attacks that basically try all possible combinations of usernames and passwords to "force" their way into your WLAN. When you make the hacker's job much longer and more difficult, there is a greater likelihood you will catch the intrusion attempt and resolve your network vulnerability.

Initialization Vectors

Repetition is bad because it makes it easier for hackers to decipher the data channel for the average LAN. Initialization vectors make the cipher stream repeat, and it is that very repetition that creates vulnerability in your WLAN.

Shared Keys

One of the methods meant for protecting your WLAN is the element that can be most easily compromised. "Shared" cipher keys by their very definition constitute a vulnerability because they can be "shared" with hackers as well as legitimate employees. The entire basis of maintaining security is highly dependent on keeping these keys secret and in the possession of authorized users only.

In the previous section we saw that hackers often try every possible username and password combination in order to try and "force" access privileges into your WLAN. Your encryption keys must be changed often, otherwise you have very little means to protect yourself against a hacker attack.

WEP uses the RC4 keys, but their deployment is poor at best due to the fact that a hacker can sometimes intercept the key just by examining the first few packets. (There are a number of other programs that do not have the same RC4 vulnerabilities; they do not leak the key schedule in each packet transmission.) Although this type of interception is often used by more advanced hackers, in fact there a number of automated means that have made this type of attack much more accessible to almost anyone interested in a simple point-and-click interface to run scripts to intercept information pertaining to your wireless network.

Checks and Balances for Packets

It is essential to maintain the privacy and substance of each packet during wireless transmission handled by cyclic redundancy checks. However,

CRC is not always sufficient to maintain the substance of the encrypted packets because it is quite possible for someone to intercept and modify the data channel. This means that these types of protection mechanisms are not sufficient to protect your WLAN from a hacker attack.

Using encryption enables you to protect yourself so that you do not become an easy target for a hacker attack. If you use protocols that do not employ encryption, you are leaving yourself open to a cryptographic attack on your WLAN.

Authentication

Accessing the network need not necessarily depend on trying to crack the access codes; it could be done by something as simple and easy as stealing the actual wireless network interface card already configured with its unique MAC address to access the wireless network.

In the vast majority of WLANs, no authentication is actually taking place. At a minimal level, only verification that the wireless device is set to use the proper SSID occurs. Systems that screen out devices based on identity are highly vulnerable because it is a simple and easy matter to "spoof" or fake the identity of your wireless device based on the SSID. Sometimes you only require just that piece of information to log into the wireless network. How secure is that?

Authenticating the device often relies on the simplest form of "shared key challenge response" mechanism. The attack most common in this type of authentication is the hacker who is between the wireless workstation and the access point using challenge response authentication mechanisms that proceed in one direction only. However, an added level of protection is possible when authentication occurs on both sides in order to verify that both the users and network are authorized to use the network resources.

Location! Location! Location!

The 802.11 standard has become enormously popular in a diverse number of implementations including hospitals, airports, retail outlets, and businesses.

The attacks, however, are growing significantly, so that having a wireless network is almost a guarantee that your private information will leak out to the hacker world. The significant risks in wireless security include:

- Privacy attacks
- Data substance and integrity
- Wireless network availability

Attack Patterns

Wireless attacks are either active or passive, as shown in Figure 4.5.

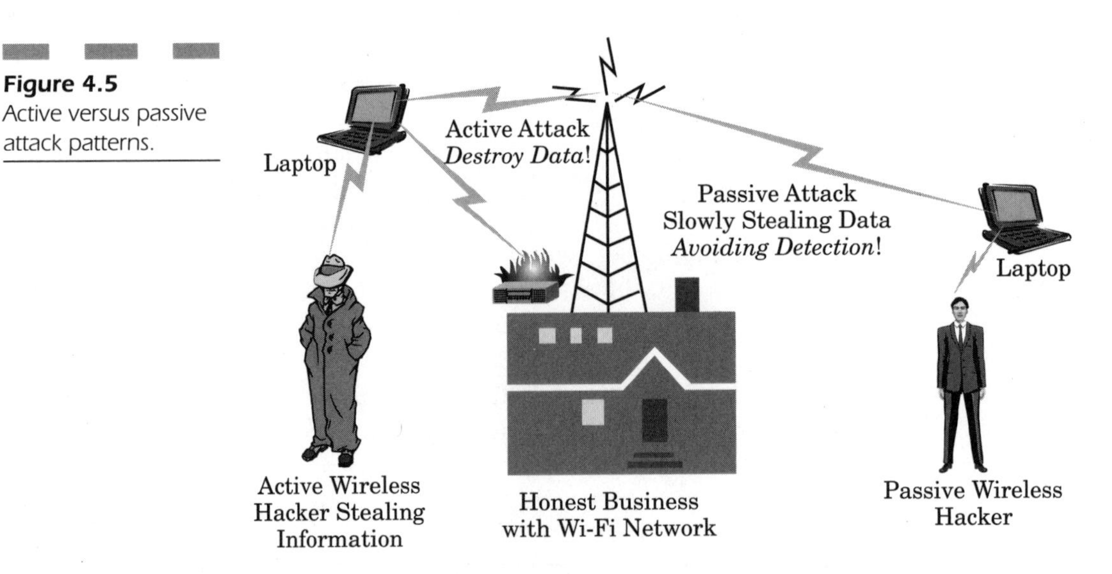

Figure 4.5 Active versus passive attack patterns.

Active Attack Patterns

An active attack constitutes a pattern where a hacker attempts to modify your data channel, messages, or files. With constant vigilance you will be able to catch this type of attack; however it is difficult to prevent this type of attack without actually pulling the plug of your WLAN.

Active attacks include: denial of service (DoS) and message alteration.

Denial of service attacks A DoS or distributed denial of service (DDoS) is an active attack pattern that prevents legitimate users from using their wireless network. There are a number of risks because these attacks prevent local and remote users from using your network resources. Besides the problems with destroying your network connectivity, you also lose business opportunities, revenue, and good public opinion.

Message alteration In this type of attack, the hacker alters the real message by either adding, erasing, or changing the sequence of the message. This removes the trust factor of your message and makes all your traffic unusable.

Passive Attacks

In these attacks, an unauthorized user acquires access to your network data sources. There is no alteration of message content, but it is possible to eavesdrop on the transmission. Passive attacks are meant not to disrupt, but to acquire information flowing across your wireless network.

Replay In this type of passive attack, the hacker intercepts or eavesdrops on your data channel. The hacker does not do anything to compromise your systems at first, but can resend altered messages to an authorized user pretending to be the system host.

Eavesdropping This is a passive attack in which the hacker listens to all your network transmissions in an effort to acquire information flowing from one wireless workstation to the access point.

Traffic analysis The hacker analyzes your traffic pattern through this type of passive attack to determine what network patterns exist. He can then use all the information acquired to gain information about the traffic from each user on your wireless network.

Conclusion

It is understandable that the nature of the wireless LAN makes it fraught with a number of wireless security risks.

Most WLAN devices come out of the box having no actual means of security to protect them against hackers. It is the responsibility of every user to ensure (as much as humanly possible) that the best possible safety precautions have been taken so that your systems are shored up against the most common problems, such as changing default values and passwords.

If you are mindful of your environment and wireless transmissions, you can effectively protect your systems against attack and ensure your WLAN is as secure as it possibly can be in the face of new hacker attacks.

CHAPTER 5

The 802.11 Standard Defined

In 1997, after seven years of work, the IEEE published 802.11, the first internationally sanctioned standard for wireless LANs. With 802.11b (2G) and 802.11a (3G) WLANs, mobile users can get Ethernet levels of performance, throughput, and availability. This chapter defines the standards-based technology that allows administrators to build networks that seamlessly combine LAN technologies to best fit their business and user needs.

1. The 802.11 standard defines two modes: infrastructure mode and ad hoc mode. In infrastructure mode, the wireless network consists of at least one access point connected to the wired network infrastructure and a set of wireless end stations. This configuration is called a *basic service set* (BSS). An *extended service set* (ESS) is a set of two or more BSSs forming a single subnetwork. Since most corporate WLANs require access to the wired LAN for services (file servers, printers, Internet links) they will operate in infrastructure mode.
2. Ad hoc mode (also called peer-to-peer mode or an Independent Basic Service Set, or IBSS) is a set of 802.11 wireless stations that communicate directly with one another without using an access point or any connection to a wired network. This mode is useful for quickly and easily setting up a wireless network anywhere that a wireless infrastructure does not exist or is not required for services, such as a hotel room, convention center, or airport, or where access to the wired network is barred (such as for consultants at a client site).

The 802.11 Standard

The evolution of the IEEE 802.11 standard for wireless local area networking (WLAN) has pushed for higher and higher data speeds with the concept of making mobile computing devices a realistic alternative to the "wired" desktop machine. Although wired LANs have been predominant for networking, wireless applications have become essential, considering the requirement to have mobile computing available for most facets of an enterprise.

Issues to Consider

When you are deciding what important issues to consider for your wireless network, it is important to take the following points into consideration. See Figure 5.1.

- Integrating your wireless network with your wired LAN
- Dealing with several access points
- Radio interference
- Implementing proper network security

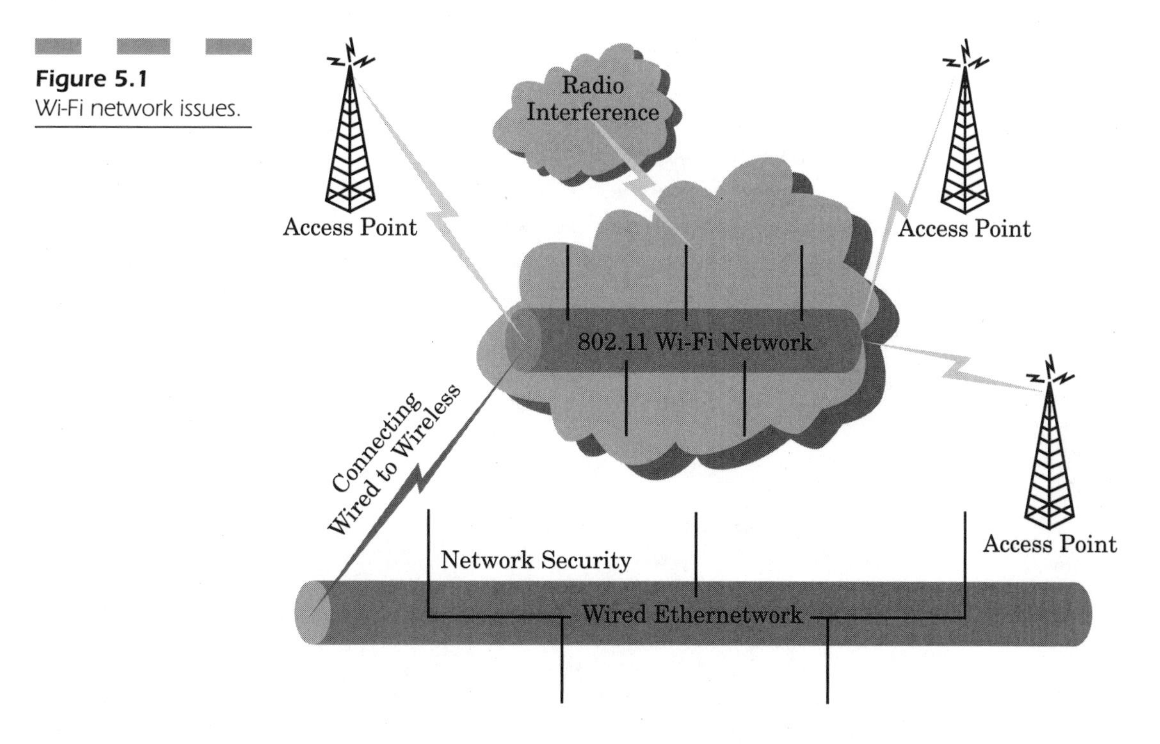

Figure 5.1
Wi-Fi network issues.

Integration is an important issue because it helps you determine how you can access all your regular LAN services through your wireless or mobile computing workstations and handheld devices.

Wireless workstations will require access to file servers, print servers, and other network resources so that users can share documents and files with other workstations on the wired LAN. When implementing a comprehensive integration strategy, all your systems will function together seamlessly so that a wired user would not even notice he is operating on the wireless network.

Dealing with several access points can become difficult. In a large wireless network, you will have several users scattered in various departments. Deploying the network can help you save IT costs, but much as in a cell phone network, you need to have capabilities so that a wireless user can literally "roam" from the range of one access point in the accounting department to another in the production area.

The 802.11b standard falls into the commercially licensed radio spectrum with many other wireless devices including Bluetooth, cordless phones, and others. The problem here is that there are a number of devices which inadvertently cause interference with this standard. The result is reduced throughput, slower connections, or broken connections. In contrast, however, the 802.11a standard falls in the 5-GHz unlicensed spectrum, so it is somewhat less common for any interference to be generated in this incarnation of the 802.11 standard.

The most important tie that binds all these elements involves network security. Since the 802.11b standard falls in the same radio spectrum as many other devices, several devices exist that can be easily modified to eavesdrop and intercept WLAN transmissions. This requires you to be more security conscious (Figure 5.2) by implementing several key elements for your WLAN:

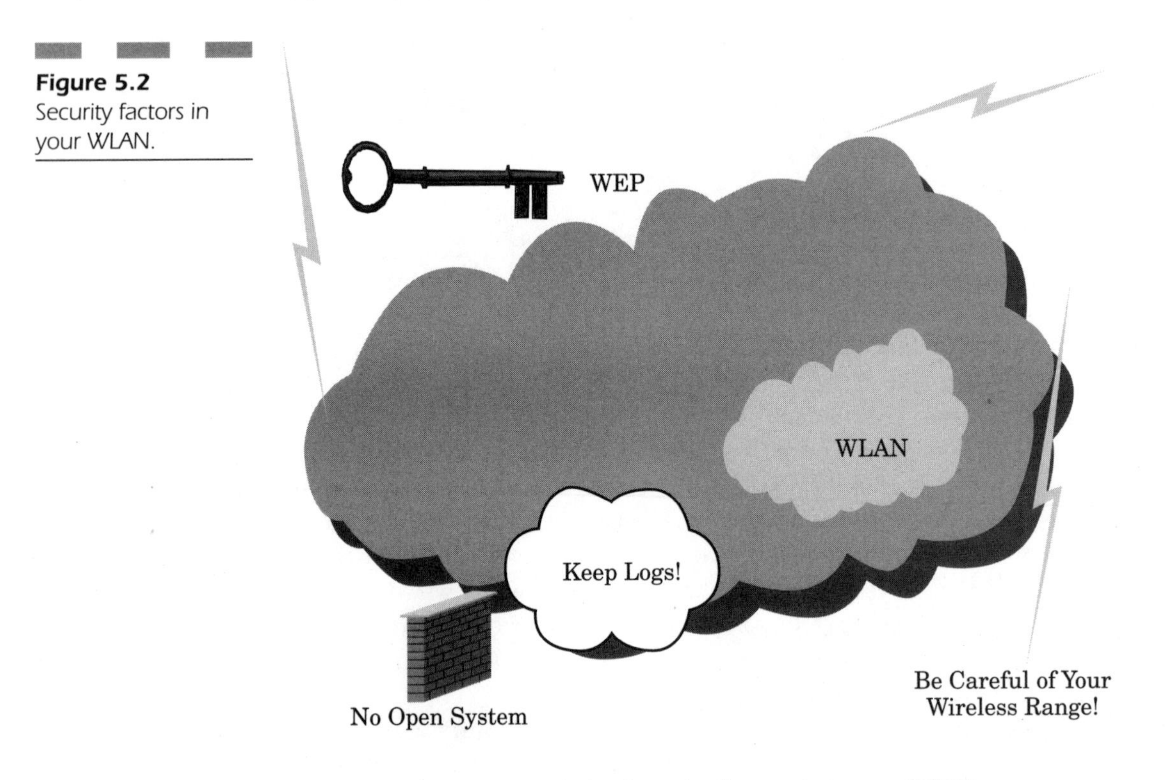

Figure 5.2 Security factors in your WLAN.

- Wireless encryption (wired equivalent privacy, or WEP)
- Do not have an "open system" that allows any wireless station to join your network; instead have each wireless network interface card's unique MAC address programmed into your access point so that only authorized wireless workstations may connect

- Be aware of the range of some of your wireless transmission devices; hackers can easily access your network from just beyond the perimeter of your building.
- Keep logs! This is your best and sometimes only defense to determine if someone is trying to attack your wireless network and gain access to mission-critical systems through your wireless link.

Expanding the Network Standard

The 802.11 standard evolved from the wired IEEE 802.3 Ethernet standard restricted within the physical (PHY) and the medium access control (MAC) sublayers.

The primary difference between physical and wireless networks is the basic service set (BSS), which is composed of at least two wireless stations or nodes (STAs) that have recognized each other and have established communications between them.

Stations are able to link directly to each other with peer-to-peer (P2P) sharing for a specific area of wireless coverage through an area that is usually called an "ad hoc" network or *independent basic service set* (IBSS).

Ad Hoc Networks

In most ad hoc networks, the BSS has at least has one access point whose primary responsibility is to create a link between the wired and wireless networks. An access point is very much like a base station used on a cell phone network to provide the most wireless coverage for associated cells in different locations. When the access point is functioning, wireless stations do not communicate on a P2P method; instead all communications between stations and the wired network are sent through the access point.

Since 802.11 access points are actually "fixed stations," they create the network infrastructure. The BSS in this setup is functioning in "infrastructure mode."

Extended Service Set

An extended service set (ESS) is composed of several BSSs, each of which contains its own access point linked through a distribution system (DS).

While a DS can actually be any type of network, it must be connected to a wired Ethernet network. Any mobile wireless workstation can roam from one access point to another through one contiguous wireless coverage area.

Wireless Radio Standard

The 802.11 standard offers two distinct types of PHY, including two RF technologies designated as:

- Direct-sequence spread spectrum (DSSS)
- Frequency-hopped spread spectrum (FHSS)

The 802.11b standard in both DSSS and FHSS PHY designations is created to satisfy FCC regulations to operate in the 2.4 GHz ISM band. This radio spectrum for the 802.11b standard is allocated differently for every section of the world. The radio spectrum is designed as follows:

TABLE 5.1 Regionally Allocated Spectrum

Country	Frequency Spectrum
United States	2.4000–2.4835 GHz
Europe	2.4000–2.4835 GHz
Japan	2.471–2.497 GHz
France	2.4465–2.4835 GHz
Spain	2.445–2.475 GHz

FHSS and DSSS PHYs support both 1 and 2 Mbps. DSSS systems employ the same type of radio transmission as GPS systems and satellite telephones. Note that each information bit is linked through an XOR function that has an increased *pseudorandom numerical* (PN) sequence, which results in a higher speed digital stream modulated on a carrier frequency through *differential phase shift keying* (DPSK).

The Standard Algorithm

To explain the 802.11 standard more fully, the period between the end of the packet transmission and the start of the ACK frame is one *short interframe space* (SIFS). The ACK frames have an increased priority over other wireless traffic. The 802.11 standard permitting fast acknowledgment is one of the most important features it offers since it requires ACKs to be supported at the MAC sublayer. Any other transmission is required to pause for at least one *DCF interframe space* (DIFS) prior to its transmission. Should the wireless transmitter detect that the medium is busy, it can then determine a random backoff time interval by setting an internal timer to a specific number of slot times.

When the DIFS expires, the time starts to decrease; when the timer approaches zero, the station can start to transmit. Should the channel be used by another station prior to the timer's approaching zero, then the timer setting is maintained at the decreased value for future transmissions. This method depends on *physical carrier sense*, in which, essentially, every wireless station listens to all the other stations on the local wireless network.

A common problem is the *hidden node*. In order to defeat this problem, a second carrier sense method called the *virtual carrier sense* permits a station to reserve the medium for a designated interval of time by using RTS/CTS frame.

When STA-1 sends an RTS frame to the access point, then the RTS is not received by STA-2. The RTS frame supports the duration/ID field that determines the period of time in which the wireless medium is reserved for future transmissions.

Reservation information is recorded within the *network allocation vector* (NAV) for all stations that are detected in the RTS transmission frame. When the RTS frame is received, the access point replies with a CTS frame composed of a duration/ID field that designated the time interval to reserve the transmission medium.

If STA-2 does not detect the RTS frame, it will detect the CTS frame and update NAV in response. This indicates that collision can then be avoided by using the nodes hidden from other wireless stations.

Note that this RTS/CTS procedure is activated with respect to the user-specified settings. It can always be used or never be used for packets that are in excess of a specific length.

DCF is the fundamental media access control method for the 802.11 standard. The *point coordination function* (PCF) is the optional extension to DC that offers time-division duplexing capabilities to deal with time bounded and connection-centered services that involve wireless transmissions.

Address Spaces

The 802.11 standard does permit different address spaces for the following disparate areas: distribution system, wireless media, and wired LAN infrastructure.

Note that the actual 802.11 standard only describes addressing when dealing with a wireless medium; however it does facilitate integration with 802.3 wired Ether-networks.

Address compatibility is maintained throughout all the different flavors of 802.11 since in most of these installations, the distribution system is an 802.11 wired LAN that has all three logical addressing spaces exactly the same leading; thus, there is little or no distinction between these areas from the point of view from anyone trying to attack your wireless LAN.

The 802.11 Standard in Security

Security is the most important element that seems lacking in many 802.11 implementations. Most people are left with the misconception that 802.11 is an insecure medium that is very vulnerable to attack. The fact is that the 802.11 standard supports two primary methods of protection: authentication and encryption.

Authentication is the mechanism used when one wireless workstation is authorized to talk to a second station in a specific wireless coverage area. Authentication is created between the access point and every station while functioning in infrastructure mode.

Authentication is either an open or a shared-key system. This means that any wireless workstation can request authentication so that the wireless workstation receiving the request may grant authentication to any request. Alternatively, it may grant authentication only to stations on a user-defined list.

In a shared-key system, only stations that have a secret encrypted key can be properly authenticated. This means that shared-key authentication is available just for systems that have the optional encryption functionality. See Figure 5.3.

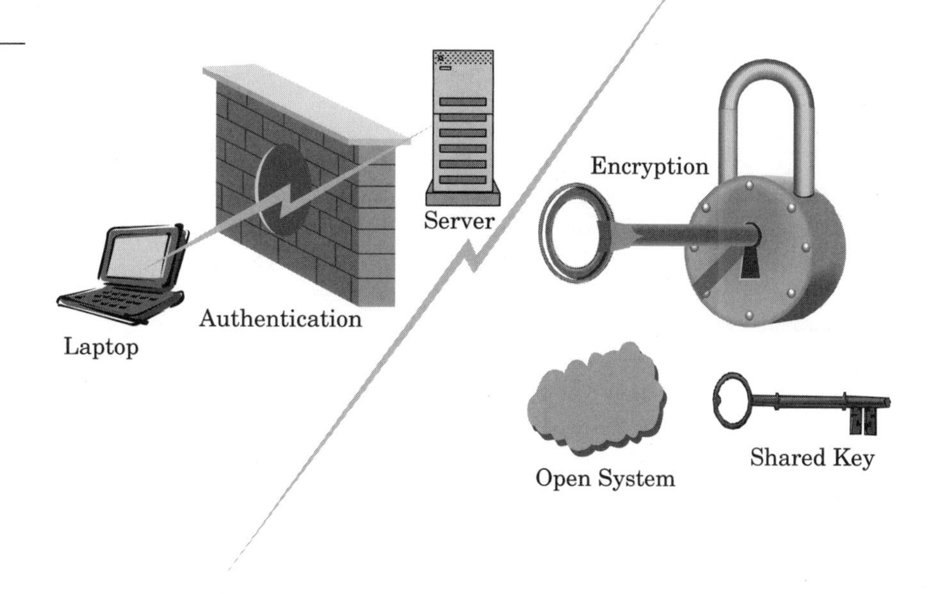

Figure 5.3 Authentication and encryption.

Encryption

You can implement wireless encryption schemes in your WLAN with the intention of offering an increased level of security analogous to what you would come to expect from sending data over a wired Ethernet LAN, commonly referred to as wired equivalent privacy (WEP). The WEP functionality employs the RC4 PRNG algorithm to provide a high level of encryption that is both strong and efficient.

Timing and Power Management

In order to achieve the most functionality from 802.11 wireless connectivity, you need to exercise control of both timing and power management. Synchronization is maintained using wireless beacons, with all station clocks within a given BSS communicating through time-stamped transmissions.

When functioning in infrastructure mode, an access point functions as the timing master to produce timing beacons. Under these conditions, synchronization is supported inside 4 microseconds in addition to propagation delay. The timing is important to keep power usage as low as possible (Figure 5.4).

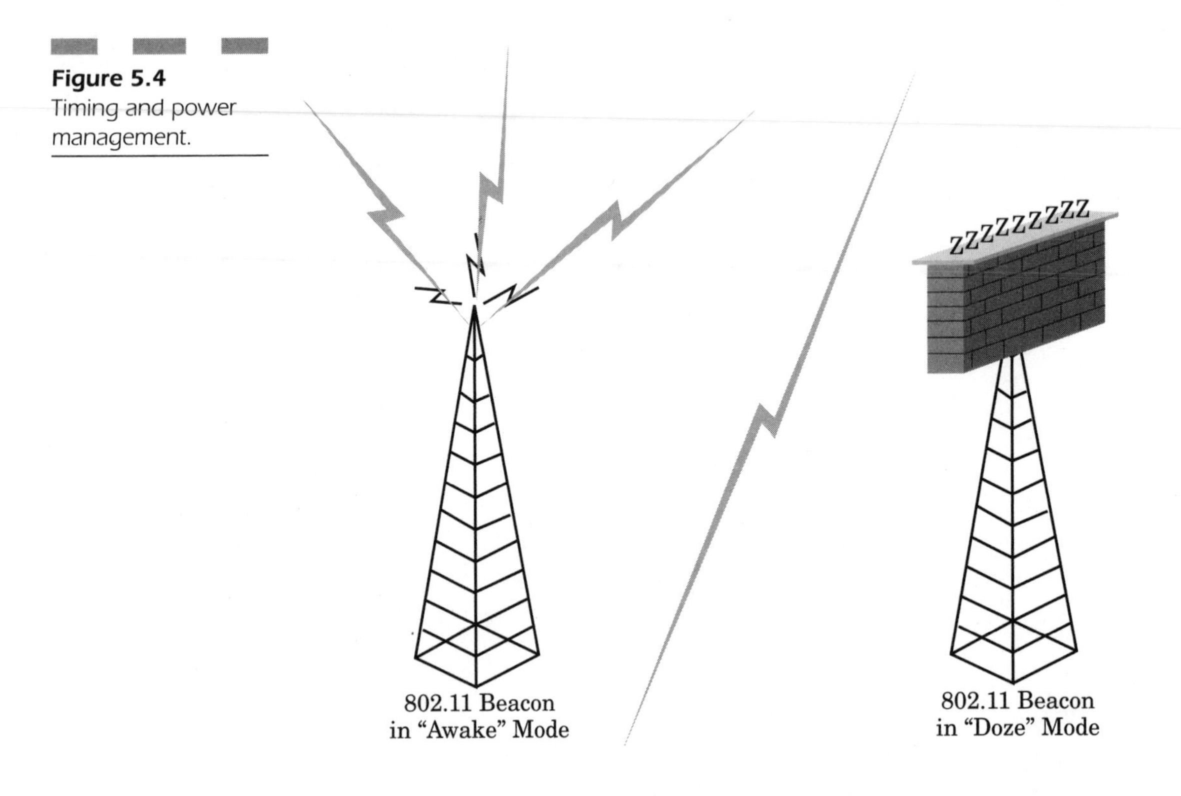

Figure 5.4 Timing and power management.

There are two primary types of power saving modes: awake and doze.

When working in "awake" mode, the wireless stations are powered on 100 percent and can receive or send packets constantly. Nodes must contact the access point prior to "dozing" off. In "doze" mode, nodes must actually come into an awake state to monitor the frequency every so often, to see if the access point has queued messages waiting for it.

Roaming in 802.11 The 802.11 wireless standard is not as defined in the 802.11 standard, but it does specify the basic message formats necessary to support roaming. Most network vendors interoperate so that wireless equipment today is not tied to any one vendor. One of the devices used to facilitate roaming is the *inter-access point protocol* (IAPP), which enhances your multivendor interoperability to the roaming capabilities of 802.11 such that roaming is possible for at least one or two ESSs.

Speed

The Wireless Ethernet Compatibility Alliance has instituted a *complementary code keying* (CCK) waveform designed to increase to DSSS speeds to 5.5 and 11 Mbps for the same bandwidth. There is also the possibility of being backward compatible so that as range increases, throughput increases accordingly.

High-speed mobile computing applications use the 802.11 standard to encrypt links between the wireless network cards and the 802.11 access points. This enables you to achieve a reasonable level of security that can be maintained while communications remain private.

Compatibility

Unlike wired Ethernet, 802.11 does not adhere to one unique standard that is compatible with all vendor devices. In corporate environments, it is necessary to use equipment that follows the 802.11 standard without any proprietary features that cause incompatibility.

The 802.11 standard uses only one MAC protocol, but exists within three physical (PHY) layers:

1. Frequency hopping (1 Mbps)
2. Direct sequence (1–2 Mbps)
3. Diffuse infrared

All these physical layers are completely distinct and incompatible with each other (Figure 5.5).

Today, 802.11 has been adopted by all vendors in this field. Its emerging specifications have essentially redefined wireless communication for wireless LANs.

The wireless physical layer standard for the common 802.11b has products that can operate at 11 Mbps. Because of high demand for greater speed for wireless networks, the 802.11a has now become a more commonly distributed standard. From a security standpoint, 802.11a operates in the higher 5-GHz band, for which there doesn't exist as much eavesdropping equipment to intercept your signal.

There are also options to extend the physical layer of the 802.11 standard with respect to enhanced security and adding quality of service (QoS). These features and functions yield increased interoperability for your WLAN.

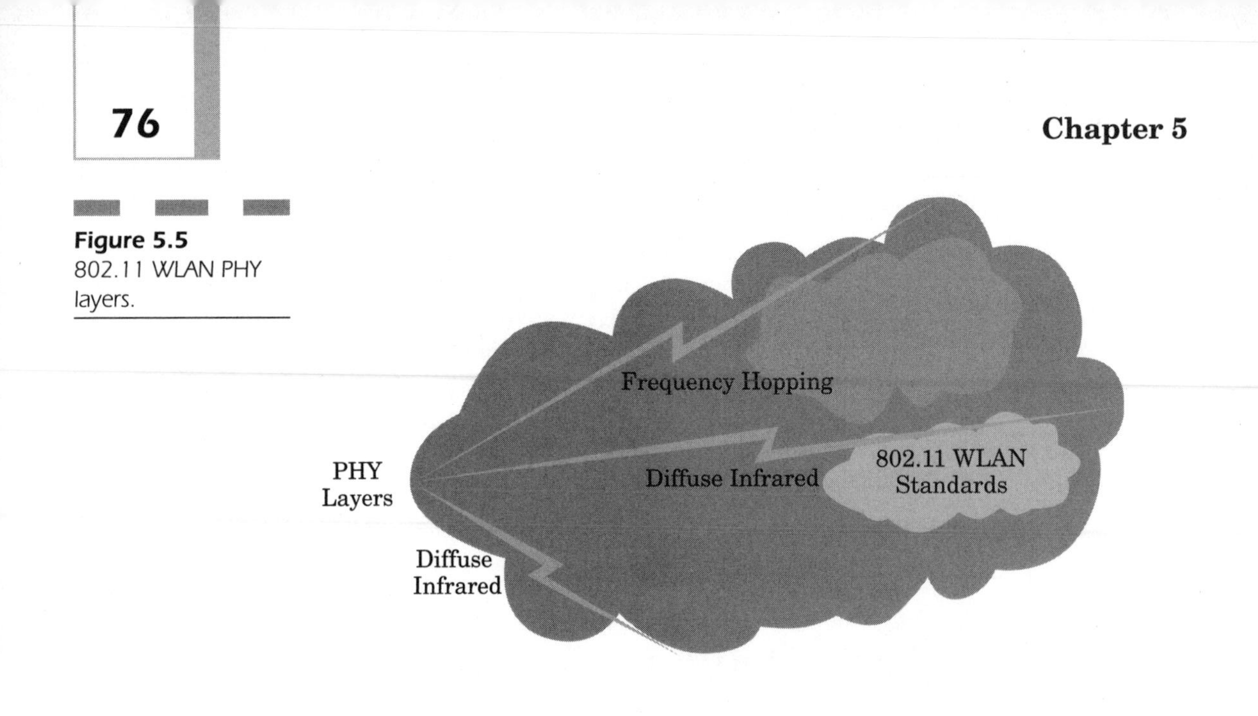

Figure 5.5
802.11 WLAN PHY layers.

Standard "Flavors" of 802.11

In order fully to understand how 802.11 has evolved, we should investigate different flavors that define how the standard can satisfy different needs and speeds.

802.11a

Because the physical layer of this specification involves the 5-GHz band, it is becoming the common replacement for the widely distributed 802.11b. It uses eight available radio channels. In some foreign counties, however, it is possible to use 12 channels. 802.11a allows for a high throughput of 54 Mbps per channel. The greatest user throughput is about half this value, because throughput is shared among all users who are currently transmitting data on a given radio channel. The data rate proportionally decreases as the distance between the user and the radio access point increases.

In the majority of implementations, the data throughput will be greater than 11 Mbps. Furthermore, with more radio channels you achieve increased protection from any hacker interference from a rogue access point.

802.11 products have become increasingly available in the latter half of 2002 with more and more vendors offering products compliant with both 802.11a and 802.11b. Prices have decreased significantly as 802.11a is quickly becoming the standard for WLANs.

802.11b

This is the most commonly used 802.11 standard. It has a physical layer standard that functions in the 2.4-GHz band, using three radio channels. The highest speed throughput link rate in this flavor is 11 Mbps for each available channel. The greatest user throughput is about half this value since the throughput is actually shared by all users working on each radio channel, whose data rate proportionally decreases as the distance between the user and the access point increases.

Your 802.11 wireless installations may experience significant constriction in maximum speed as the number of active users increases. However, the limit of using three radio channels may cause interference with other access points within your WLAN.

802.11d

802.11d is supplementary to the media access control (MAC) layer in 802.11 to promote global use of 802.11 WLANs. Its basic premise is to provide access points with the ability to communicate information on available radio channels with sufficient user device power levels for maintaining good signal quality while at the same time conserving energy.

The 802.11 standard cannot legally operate in some countries; therefore the purpose of 802.11d is to add extra features and restrictions that permit wireless networks to function within the rules of foreign territories.

When dealing with countries whose physical layer radio requirements are different from those of the United States, the 802.11 WLAN is inapplicable. Due to these problems, equipment vendors do not wish to produce equipment usable in foreign territories since there would be so many different specifications it would be impossible to make a profit by building custom country-specific products.

The most difficult problem is that users cannot roam around the world and still expect their wireless NIC cards to function. The only solution in such cases is to build a method to inexpensively flash to the cards firmware that takes advantage of the unique requirements of the country the hardware is shipped to.

802.11e

The 802.11 has physical standards of a, b, and g that provide supplementary QoS support to the MAC layer for your LAN applications. This is provided for service classes with managed levels of QoS for the following applications: data, voice, and video.

802.11e provides useful features and functionality for making a distinction between various data streams. WLAN manufacturers use QoS as a feature as a distinction in their products, but the down side is that many elements are still proprietary until the standard is set.

These products will only be successful when the 802.11e standard becomes more defined and products start to roll out in early 2003. However, the prices of these initial product offerings won't become reasonable until late 2003 or even early 2004.

802.11f

The idea of this standard is to achieve interoperability among several WLAN network vendors and manufacturers. This standard determines the access point registration within a network. It also covers the exchange of information from one access point to another when a user migrates from one cell to another (as in a cell phone network).

802.11g

The 802.11g standard uses *orthogonal frequency division multiplexing* (OFDM) manipulation; however, for backward compatibility, it can also work with the more commonly used 802.11b devices by supporting complementary code keying (CCK) and packet binary convolutional coding (PBCC) modulation.

802.11g offers speeds in the same range as 802.11a as well as backward compatibility; however the modulation issues include unresolved problems between key vendors whose support is divided between ODFM and PBCC modulation schemes.

The ultimate compromise is the adoption of support for 802.11b's CCK modules so that it will ultimately support all three types of modulation. The advantage is that vendors can have dual mode devices that function in both 2.4 GHz and 5 GHz and use OFDM for both modes to cut costs.

This means that 802.11g could theoretically excel in the European areas should 802.11h not succeed as the high-speed standard in that part of the world.

802.11h

Thiscompeting standard is trying to satisfy European power regulations for transmission in the 5-GHz band. These products must have *transmission power control* (TPC) as well as *dynamic frequency selection* (DFS).

TPC restricts the transmission power to the least amount necessary to reach the user who is farthest away. DFS then chooses the radio channel at the access point to reduce interference with other networked systems functioning in the same radio portion of the spectrum.

Its competition with 802.11 increases its European acceptability for 5-GHz WLAN products. The actual acceptance of products that use 5 GHz with TPC and DFS won't officially take place until the latter half of 2003 and perhaps even as late as early 2004.

802.11i

802.11i is a key element to improving MAC layer security and is applicable as an alternative to WEP applications. Most manufacturers ship products without setting any security features. The products come out of the box unsecure, without encryption, and most users have no idea how to implement the most basic security measures.

802.11 specifies a portion of the security features that must support solutions that begin with firmware upgrades that can only be accomplished using the *temporal key integrity protocol* (TKIP) in combination with and advanced encryption standard (AES) (iterated block ciphers) and TIKP backwards compatibility.

For WLAN products to achieve Wi-Fi certification, they must implement additional security features above and beyond those already set in the standard. The constantly evolving corporate networks must integrate standard forms of encrypted modulation techniques that provide a greater level of inherent security during wireless transmissions.

Conclusion: Evolution of the 802.11 Standard

The 802.11 standard has evolved considerably and continues to be refined. One of the most common misconceptions is that this standard does not provide any significant level of protection, security, or privacy in a wireless medium. Nothing could be further from the truth. Although when you take devices like wireless access points or routers out of the box, they are designed to function in an "open system," where any wireless workstation in range can join, if you follow the specification of the standard permitting encryption and selective access control lists, 802.11 can provide a level of protection analogous to that of a wired network.

802.11b is the most commonly deployed wireless network standard today. It provides 11 Mbps of throughput, which is just barely adequate for today's hungry bandwidth intensive network applications. The 802.11 standard is vulnerable to eavesdropping because it functions in the same portion of the radio spectrum as cordless telephones and other devices. This means it is a relatively simple matter to find a listening device.

However, 802.11a is coming of age. It is in wider use as the equipment for this flavor of the standard is also being produced by most major manufacturers with backward compatibility with 802.11b. Since the 5-GHz band is unlicensed for many radio applications, it is far more difficult to design an eavesdropping device, but not impossible.

In essence, as the 802.11 standard evolves to offer greater speed it carries a greater security risk if several key options are not configured. If you diagram your wireless connection between the mobile workstation and the access point, ensure that the channel is encrypted so that if anyone does try to listen in, they only get garbage. You might also want to consider using a virtual private network (VPN) to add a further layer of encryption. The only downside to doing so is that you add a far greater level of overhead that slows down your connection. This is why 802.11a will be the dominant protocol in the very near future, once its prices drop, for it offers two primary advantages:

1. **Faster connection**—Up to 54 Mbps to effectively deal with the overhead of bandwidth-intensive applications
2. **Operation in the 5 GHz band**—There is a far smaller chance of interference from other devices functioning in the same radio spectrum.

As 802.11 continues to evolve, we will ultimately see manufacturers producing wireless LANs that can operate globally using all the slightly different wireless standards and varying frequencies. 802.11 has such potential that a universal standard is only a few years away, but security will always remain a prominent concern for users who need to configure it appropriately so that information remains both secure and private on any wireless network.

CHAPTER 6

802.11 Security Infrastructure

This chapter describes the internal workings of 802.11 and how it provides for both MAC-layer access control and encryption mechanisms, which are known collectively as wired equivalent privacy (WEP), with the objective of providing wireless LANs with security equivalent to that of their wired counterparts. This chapter also describes how the access control and the ESSID (also known as a WLAN service area ID) is programmed into each access point and is required knowledge in order for a wireless client to associate with an access point. In addition, there is provision for a table of MAC addresses called an *access control list* (ACL) to be included in the access point, restricting access to clients whose MAC addresses are on the list.

Point-to-Point Wireless Application Security

When you think of security, it is important to conceptualize how to maintain a secure connection from the point of the user to the point of the server. The first action in creating a secure wireless infrastructure is to focus on secure remote access.

Secure remote access means creating secure communication so that you can exchange various identifying items securely, including passwords, cryptographic keys, session keys, and challenge-response dialogs (Figure 6.1).

Wireless environments are vulnerable to the same types of denial of service (DoS) and flooding attacks as wired networks. Hackers try to obtain access to your internal wireless infrastructure to initiate these attacks that essentially make your wireless infrastructure collapse so that you are not able to serve your users effectively.

Point of Interception

If someone does try to compromise the infrastructure of your wireless network, you can be sure that the point of interception will take place at the location where signals are transmitted from your internal network. Any form of interception or eavesdropping is not only easy to do nowadays, but can be done with prefabricated scanners designed to pick up the transmissions of your network. Most people believe that if their

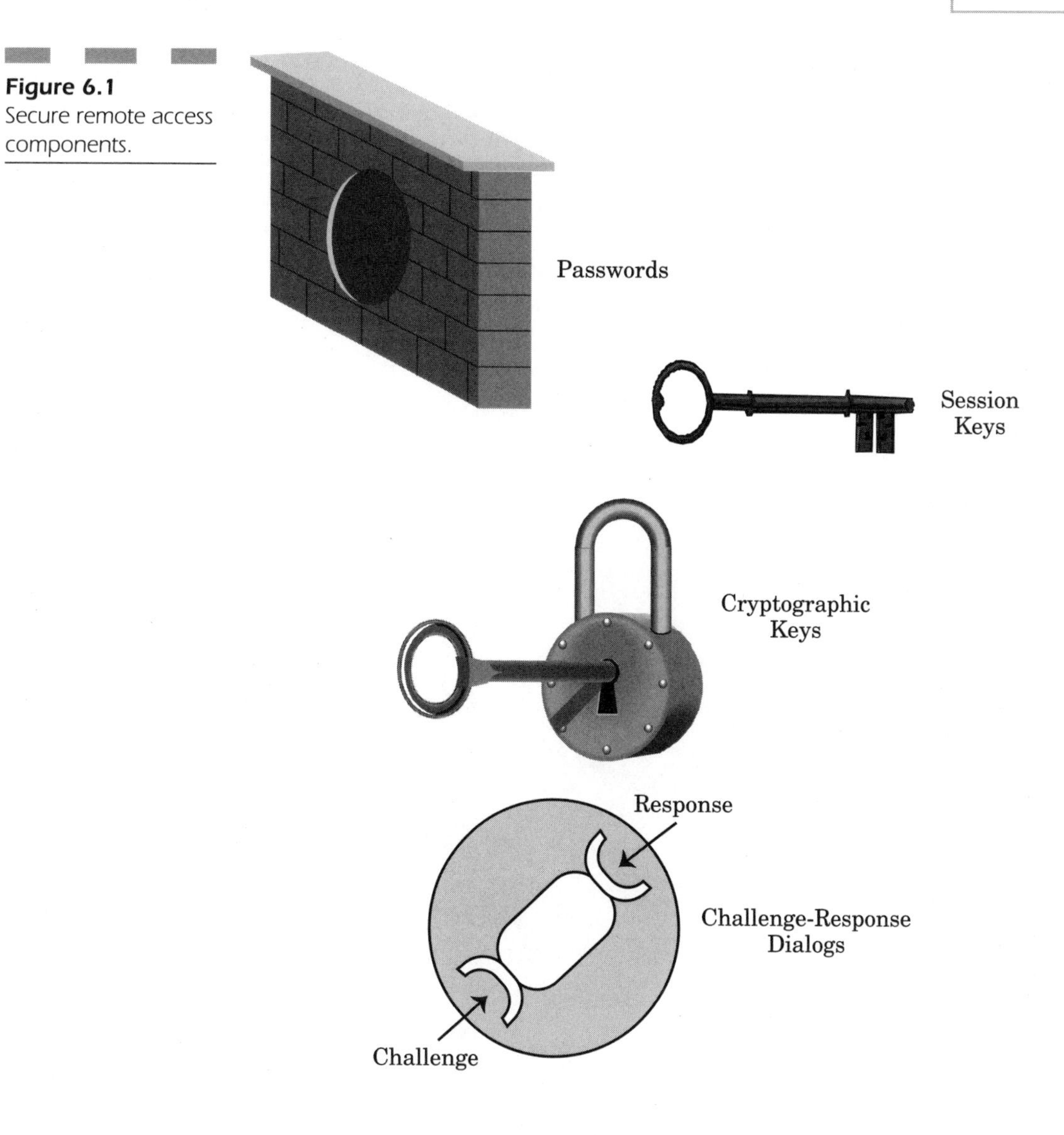

Figure 6.1 Secure remote access components.

wireless network is digital instead of analog, in some way they have a greater level of protection. Unfortunately, nothing is further from the truth. Scanners are designed now to pick up signals from either an analog or a digital environment. These scanners are sometimes not even very complicated and are very inexpensive. This makes the realm of hacking accessible to almost anyone interested in attempting an attack on your wireless infrastructure.

Wireless Vulnerability

As convenient as wireless networks are, their infrastructure is always vulnerable to attack. In fact, wireless systems throughout history have been vulnerable to electronic warfare. If a hacker is going to attack your network, wireless methods are the easiest and surest means to disrupt an entire company (Figure 6.2).

Electronic warfare and its control is divided into three primary areas:

1. Electronic counter measures (ECM)
2. Electronic support measures (ESM)
3. Electromagnetic counter-countermeasures (ECCM)

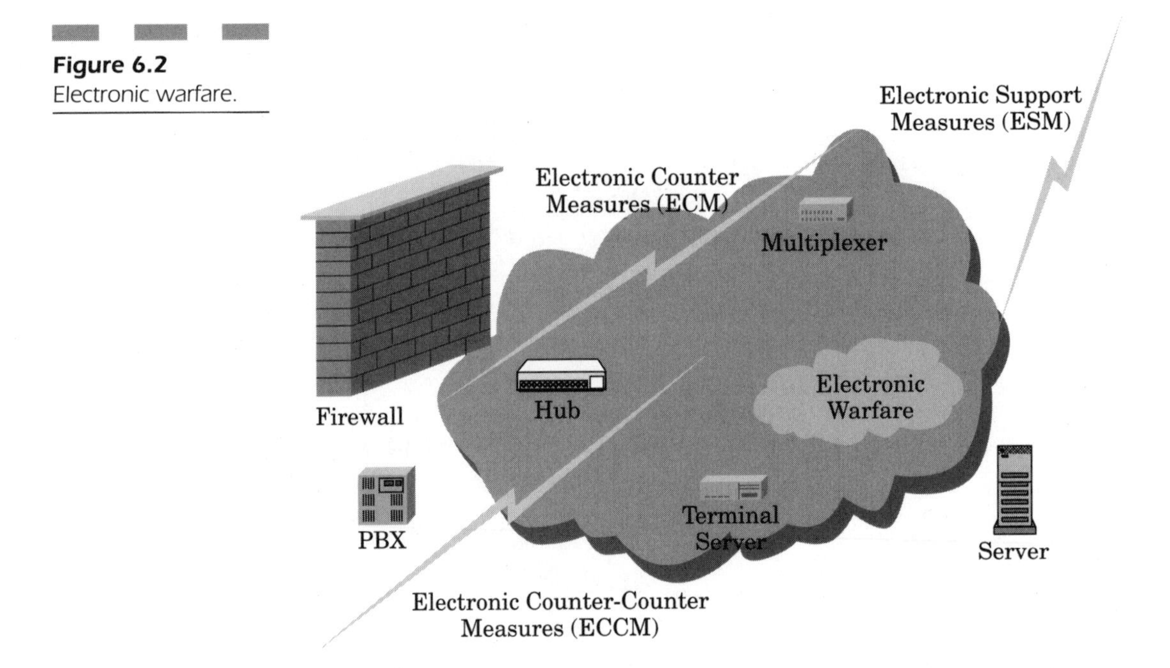

Figure 6.2 Electronic warfare.

ECMs are the actions you need to execute to stop a hacker from using your radio spectrum and causing problems with your ability to keep your wireless infrastructure intact. These types of attacks are often in the form of *jamming*, an intentional transmission of radio waves that causes serious problems in the functioning of any wireless networking device. Deception, however, is worse, since it is the manipulation of your network with the intent of misleading networking devices so that they think the hacker is actually part of your corporate network. This form of simulation is analogous to a "spoofing" attack that can promote hostile communication and lead to the intentional leakage of mission-critical data through no fault of the user.

ESMs involve the interception, identification, analysis, and localization of hackers disrupting your transmission sources. They also enable you to determine what steps you need to take to deploy the correct amount of force to counter any specified threat.

Hackers often spend an inordinate amount of time collecting intelligence for the purpose of deciphering electromagnetic data radiated by your network.

Communications intelligence, non-communications electronic intelligence (ELINT), and electromagnetic data are all part of a method that provides signal intelligence (SIGINT).

Electronic counter-countermeasures are steps you can take to protect your wireless network against future attacks. One way that these countermeasures can be used is to design your WLAN so that you are operating in ways that the hacker won't anticipate.

Moving on up! (to the 5-GHz band) One good method of staying ahead of hackers attempting to compromise your WLAN is to migrate to 802.11a so that the frequency allocation you use for your wireless transmissions is in the 5-GHz band as opposed to the 2.4-GHz band. Most readily available scanners are in the 2.4-GHz band and since the higher band frequencies are as yet unallocated for many commercial applications, it is just that much more difficult for someone to attempt an attack on your WLAN. Sometimes that is all it takes, just to migrate to a newer application of an existing technology, to put yourself one step ahead of hackers attempting to compromise your wireless infrastructure.

Fortress of solitude (wirelessly speaking) Another way of instituting wireless electronic counter-countermeasures is to isolate the building that houses your WLAN from radio frequency interference. More to the point, this means interference caused by a wireless hacker attempting to disrupt your wireless infrastructure (Figure 6.3).

Most frequencies in the 2.4-GHz and 5-GHz bands penetrate most standard building materials, but adding shielding will hamper the migration of those frequencies through your corporate facilities to the outside world. Additionally, some building materials and woods are being used in modern cell phone devices to protect users against stray RF energy released during the course of a normal telephone call. Essen-

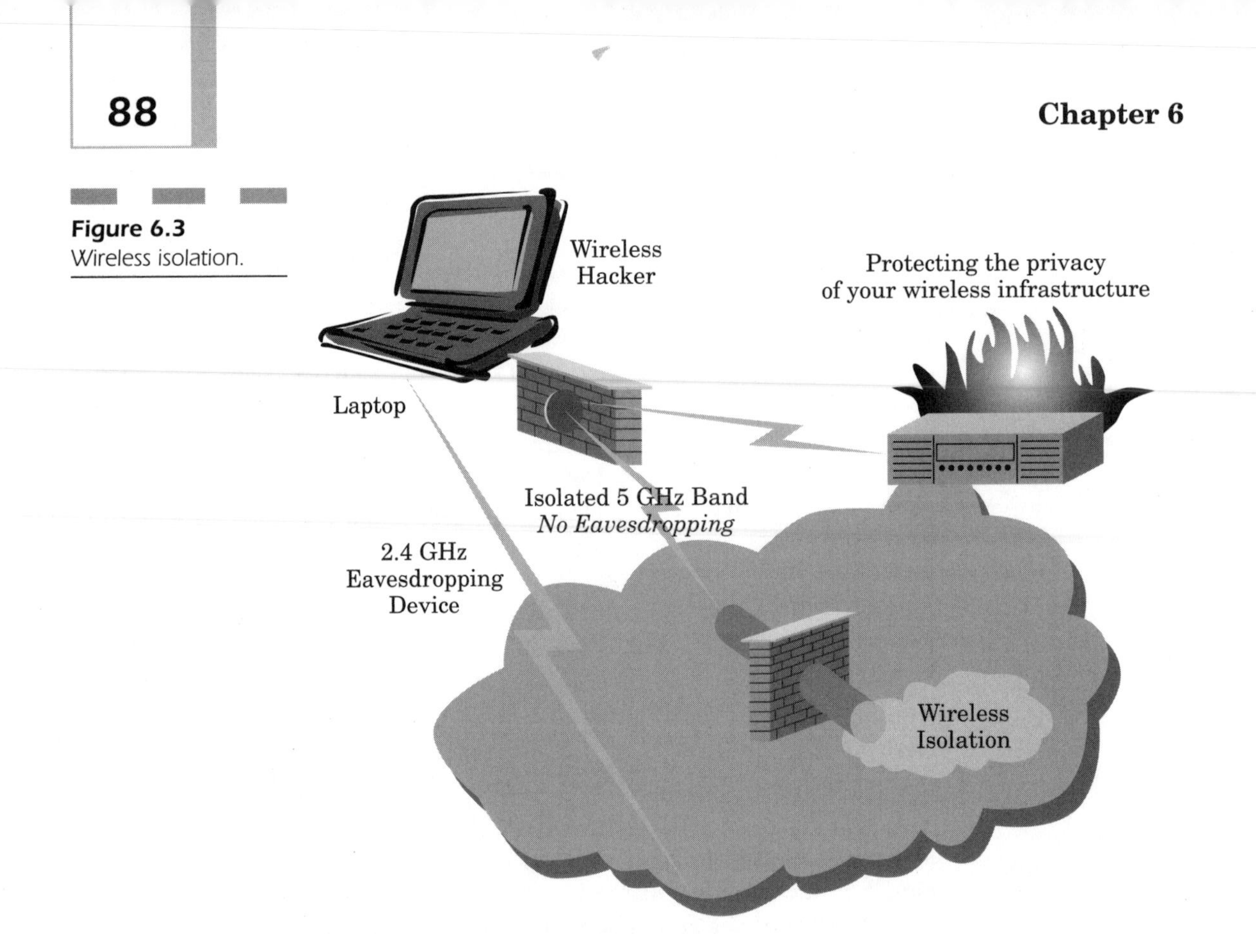

Figure 6.3
Wireless isolation.

tially, this means that you can place aluminum panels in the walls near your wireless access point to prevent the transmission of your wireless network beyond a certain distance. This means that only wireless users in your immediate corporate facilities can access the WLAN, while hackers will have a much harder time doing so. In many cases, creating this "fortress of solitude" makes it just that much more difficult for outsiders to attempt any form of electronic warfare on your system; that may be all it takes to protect yourself against the majority of hackers eager to disrupt your wireless infrastructure.

Building a Private Wireless Infrastructure

The 802.11 standard has been broken, even with all the security measures built into it, usually because people put these systems into their companies without understanding how to use the integrated security measures to protect their wireless infrastructure against attack.

In contrast, your wired infrastructure is more secure, because someone has to acquire physical access to the actual Ethernet wire in order to bypass the firewall in your organization and gain access to any system within your network.

When dealing with a wireless system, a potential hacker must get close enough to access the wireless carrier signal of your wireless access point. Most potential hackers must get within several hundred feet, but new wireless NIC cards have an external antenna designed to gain access to the network from even farther away.

Vulnerable Encryption

The Wi-Fi 802.11b infrastructure has difficulties with its encryption scheme, which can easily be decrypted. One of the ways that wireless users can make their wireless connection more secure is to connect through a virtual private network (VPN) that can be established through the wireless connection. Unfortunately, most users are either unaware of this capability or unwilling to implement it. The primary reason people are not impressed with using these forms of encryption is because they add a great deal of overhead to the connection. Encryption essentially slows down the speed of the wireless connection. In 802.11a environments this is not so bad because such environments have a maximum speed of 54 Mbps. However, since 802.11b is limited to a speed of 11 Mbps, adding encryption slows down the connection to the point of disrupting the user's wireless network connectivity.

Commercial Security Infrastructure

Many commercial companies have implemented wireless devices that permit stores to establish additional point-of-sale machines quickly when they add more departmental locations.

Wi-Fi devices are very convenient and allow these devices to work quickly without the expense of adding additional wiring to link them. However, the problem is that these devices transmit credit card numbers over the wireless network. Many of these systems were instituted without any encryption schemes; those that were used the lower 40-bit encryption scheme. Many hackers were able to intercept and eavesdrop on these signals to pick up the credit card numbers and exploit them for fraud.

Other commercial stores use wireless video cameras that transmit images over either the 2.4 GHz spectrum or a short-range wireless network so that the managers can keep tabs on all the unsecured areas of the store to prevent shoplifting. Unfortunately, hackers found this out and were able to tap into these systems to determine when store aisles were vacant so that they could direct their cohorts in stealing items from the store.

When it became clear that wireless point-of-sale machines and cameras actually were more of a security risk than a benefit, these types of wireless devices were discontinued in many areas, but not all. Hackers are eager to search for companies that still use wireless cash registers and cameras; they can then turn these items against the stores that implemented them.

Building a Private Infrastructure

In the majority of cases, when companies build their wireless infrastructure they often fail to account for privacy concerns. Security is often an afterthought, and by then it is a simple and easy matter for someone with a laptop and wireless NIC to use freely available software to roam directly onto your wireless network and have almost unlimited access to your entire intranet.

Wireless users are more sophisticated as they look for ways to compromise the privacy of your wireless network. The most common tools include "sniffers" that can listen to the network to get user passwords and steal confidential documents transmitted directly from your e-mail server. These actions are no less than corporate espionage. The most common attack is from people who understand the building blocks of your network and sit just outside your building, roam onto your wireless network from their cars, and record all your network activity.

Items to Compromise

What does a hacker look for when he monitors your network? In the majority of cases, he looks for information that he can use, sell, or modify for his own purposes as shown Figure 6.4. These include:

- Credit card numbers
- Passwords

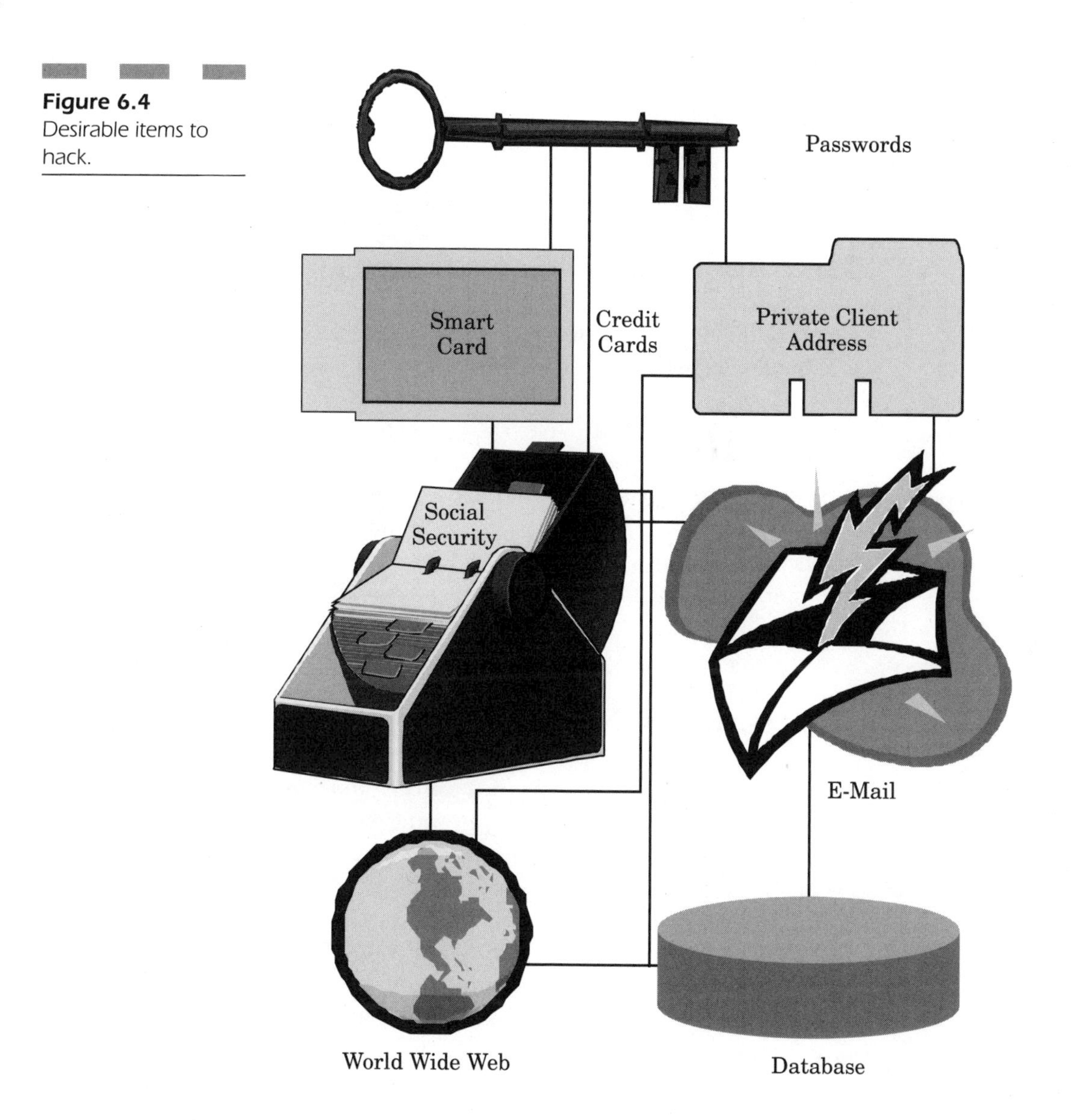

Figure 6.4 Desirable items to hack.

- Documents
- Social security numbers
- Incoming and outgoing e-mail
- Private/internal Web sites on your intranet
- Any file on your server that is accessible within your intranet

Unsecure access points are the most vulnerable areas, and they are most often attacked. This is the one element of your wireless infrastructure that is most often configured improperly. Attacks on various access points are mounted so that private information transmitted over your wireless infrastructure can be acquired.

Deploying Your Wireless Infrastructure

When deploying your infrastructure, some of the very first items you will have to consider are the following:

- The version of 802.11 to choose, (a) or (b)
- The choice of a wireless vendor
- Dealing with the security and privacy concerns of 802.11

All these issues are important when creating a WLAN, but you need to understand and analyze your overall network to determine how your final deployment will satisfy the current *and* future needs of your users.

Determining Requirements

In order to satisfy needs, your first concern involves determining specific requirements. When you start planning your wireless infrastructure, define specific requirements by first performing a key analysis of your needs. Your goal is to define what your WLAN is going to do prior to taking the next step.

Avoid purchasing or installing your WLAN without adequate planning, which begins with determining your wireless infrastructure. Without sufficient planning, your final wireless infrastructure will not satisfy your users' needs. It is important not to install the network and then have disgruntled users point out all the requirements your wireless deployment lacks.

The main requirements for planning your wireless infrastructure include:

- Immediate user needs
- Planning for future user needs
- Company needs and growth

The requirements for your wireless LAN include:

- Wireless range
- Throughput
- Security and privacy
- Battery life
- Application software
- Operating systems
- User hardware

Note that some of these requirements are different from and more complicated than what you might have planned for your traditional wired networks. This means you must really understand the issues involved in creating your wireless infrastructure.

Choosing a Flavor of 802.11

One of the more important decisions you will need to make when deploying your wireless infrastructure involves choosing either 802.11a or 802.11b. You get more speed with 802.11a, but 802.11b is much less expensive and much more commonly available. However, if you want to deploy a wireless infrastructure that is going to last for a long time, you may find it much easier to deploy 802.11a. While 802.11a is more expensive at this time, costs are going down for wireless NIC cards and access points. Most important, you gain a significant speed increase from 11 to 54 Mbps. In today's information world, you will require more speed in your wireless infrastructure for multimedia network applications that require more bandwidth.

In order to deploy the most effective solution possible, make certain you understand what capabilities you will need in your wireless structure both today and tomorrow. You can overcome 802.11 security limitations by determining your requirements. One of the factors that makes 802.11 more secure is that it functions in the 5-GHz band. This frequency spectrum is significantly different from the 802.11b use of the 2.4-GHz band, and it is much harder to eavesdrop on the signal with off-the-shelf listening equipment.

Defining your requirements is necessary for you to determine what your WLAN is going to provide your users. It is vital that you completely define your wireless infrastructure requirements, or your WLAN won't satisfy your user's needs (Figure 6.5).

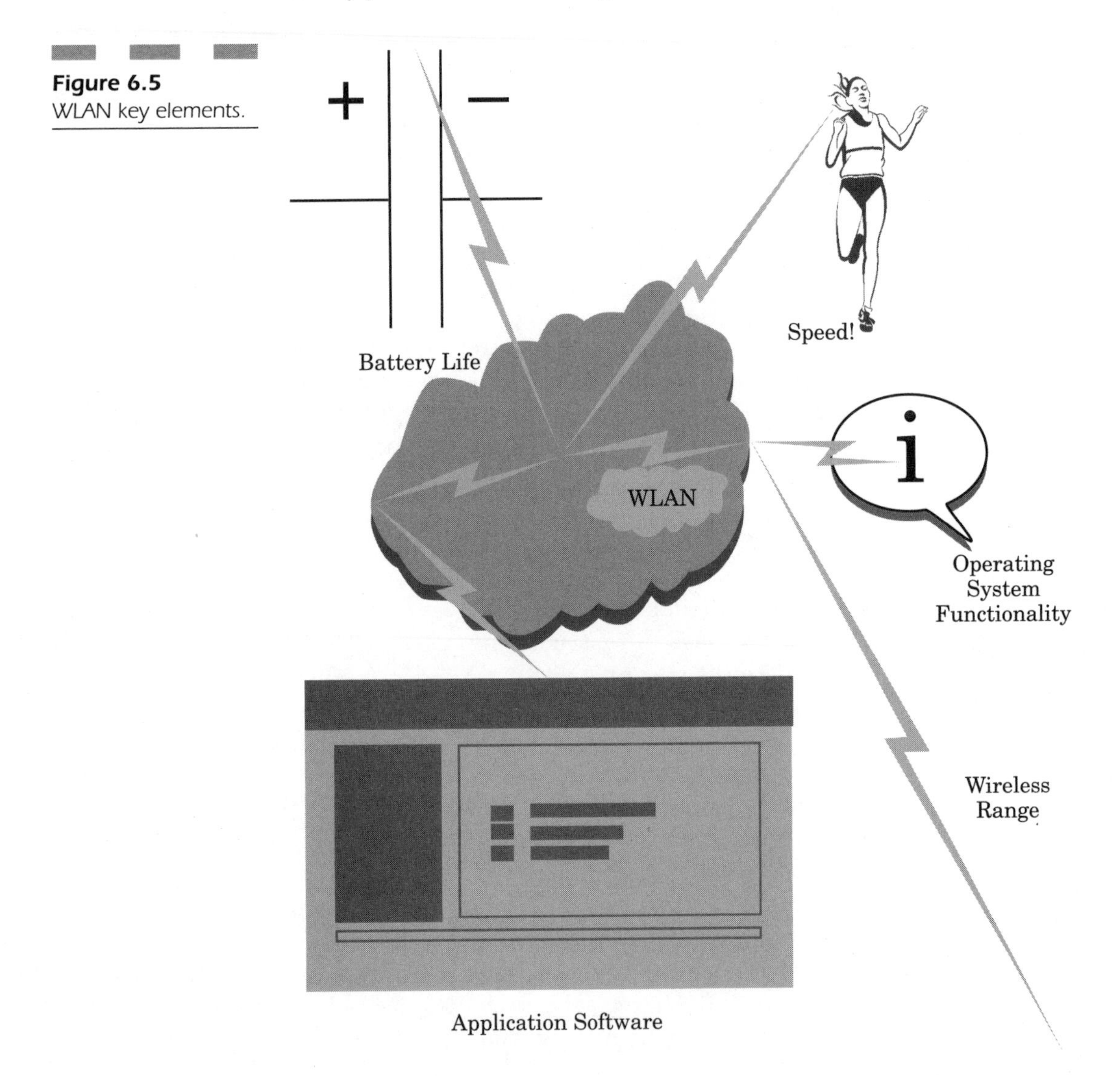

Figure 6.5
WLAN key elements.

The key requirements you need to consider include the following:

- Wireless range
- Speed and throughput
- Security
- Application software
- Battery life
- Operating system functionality

In order to understand these requirements it is important that you determine what applications you will be running on your network. Many network applications are bandwidth intensive and have increased throughput requirements. Next, you need to determine how many users will be concurrently using these applications. Understanding that throughput decreases proportionally with increased distance from the access point, you also must consider the range your users will need in order to work efficiently over your WLAN.

When you define network requirements, you should consider how many users you are planning for in your WLAN. If you are dealing with only a few dozen workers, it may only take a few hours to determine the necessary requirements. When working on larger projects with several thousand people using a WLAN in a large corporate area, you may need to invest several weeks to survey your users so that you can determine the most appropriate mechanism for your new wireless infrastructure.

It is important for you to be able to plan your wireless infrastructure with enough room for future improvement so you can meet the increasing needs of your corporate users.

When you have determined a fixed set of requirements, then you can concentrate on effectively designing your infrastructure to meet your requirements at the lowest possible cost.

The most important elements that help you reduce the cost to deploy your wireless infrastructure include:

- Choosing a vendor
- Assigning the most effective access point locations
- Designating non-conflicting access point channels
- Determining how to assign security mechanisms to protect your network
- Determining components to meet wireless infrastructure requirements
- Assigning the most efficient wireless configuration

You may find it most appropriate to create a design diagram that describes the specific configuration and components needed to meet your wireless design requirements securely. You design specifications will define how best to plan your wireless devices for secure, optimal reception. You can achieve the best reception by placing the antennas for your access points at higher elevations to get the most range. From a security standpoint, shielding your walls from stray signals from your access points helps you contain your WLAN so that people cannot hack into it or gain access to resources they are not authorized to utilize.

Security Design

Security is the most important concern in developing these requirements for your wireless infrastructure. As requirements change and networking improves in step with the evolution from 802.11b to 802.11a and beyond, understanding the dynamics of providing a secure access conduit is essential to providing speed tempered with access for authorized personnel only.

When creating your wireless infrastructure, by default, systems are designed to be "open" so that any wireless station in range of the transmitter can "roam" right onto your network. From a security standpoint this is dangerous because someone could easily try to access your system from the parking lot of your building.

You can design your system with wireless routers and access points that are easily configured to accept only transmissions from wireless stations that have been preauthorized to join your network.

Just as the dynamic host confiuration protocol (DHCP) server in a wired network assigns a static IP address to a specific workstation, wireless LANs can be configured in much the same way. The configuration dialog in most products permits an administrator to enter into the memory of the router the MAC address (a unique identifier for each wired or wireless network interface card) of each card. This means that only those stations flagged for access can roam onto the network. Any station that has *not* been authorized will not be able to join the system.

This leaves the vulnerability to eavesdropping still a problem for most wireless infrastructures. In the 802.11b framework, the 2.4-GHz frequency spread is common enough that almost anyone can get a device to eavesdrop on the signal. However, since 802.11a operates in the unlicensed portions of the 5-GHz band, eavesdropping in that frequency range is much more difficult.

Nevertheless, the question of preventing eavesdropping in the 802.11b area is the most common problem. What users can do is create a virtual private network (VPN) to mission-critical network resources when connecting wirelessly. In combination with the default level of wireless encryption, the VPN will add another layer of encryption, making it difficult if not impossible for a hacker to eavesdrop on the signal. If he were to decipher your wireless encryption scheme, then there would still be another level of decryption necessary before viewing any of the information in the wireless stream.

Monitoring Activity

One of the best tools to use to maintain your wireless infrastructure security is not any tool, but actual human intervention. The best way to defend your network infrastructure from attack is to have an actual person review the access logs and access attempts into your WLAN. If it appears that someone is gaining access to network resources at off hours or is attempting to break a password, you will be able to determine this in a relatively short period of time.

Once you can determine if someone is attempting to gain access to your systems, you can use techniques to triangulate the signal of the person attempting to break into your network. For example, you can trace the signal back to an attacker sitting in his car right outside your building, and the police can make an arrest.

There are even law enforcement agencies who can take the uncorrupted access logs from your access point and use that information as a vehicle for prosecuting would-be attackers on your system. The reason I say “uncorrupted” is because logs can be rewritten or modified by intruders so that the information is inconclusive and cannot be used against someone for prosecution. This is why early detection is the most important element in making certain that your wireless infrastructure remains secure and private.

Conclusion: Maintaining a Secure Infrastructure

In this chapter we have seen how to build an infrastructure, but most important, why it is necessary to take your time in planning every facet of your WLAN by considering your requirements. The essential ingredient in building a secure wireless infrastructure is to ensure that your users' requirements are going to be met in the short term as well as the long term.

Requirements planning allows you to make certain that when you build your WLAN, you can increase your user base and wireless infrastructure without compromising your security.

Knowing how your WLAN will increase as your building facilities and bandwidth needs grow is an important part of being able to determine how you can best bring the network to the far corners of your growing business while making certain that the security of those connections does not become compromised by anyone trying to steal information or corrupt the data on your internal network.

In the end, keeping a watchful eye on your network activity is the most important part of making certain that your wireless resources remain secure. An intrusion detection system, firewalls, and anti-virus software are all important tools that "assist" you in keeping private network resources secure, but they are no substitute for actually reviewing the access attempts and authentication entries from the logs of your wireless routers and access points.

The most difficult step is actually knowing you are under attack. Hackers usually take their time over a period of several days, weeks, or months to try and break into your system undetected. If you are watching for slow but sure activity on your log entries that indicate an attack, you can take a proactive step in assuring that your information remains secure and private.

Your final objective will be to make certain you can create the most secure wireless infrastructure possible in an effort to protect your entire internal network against corporate espionage, damage, and attack.

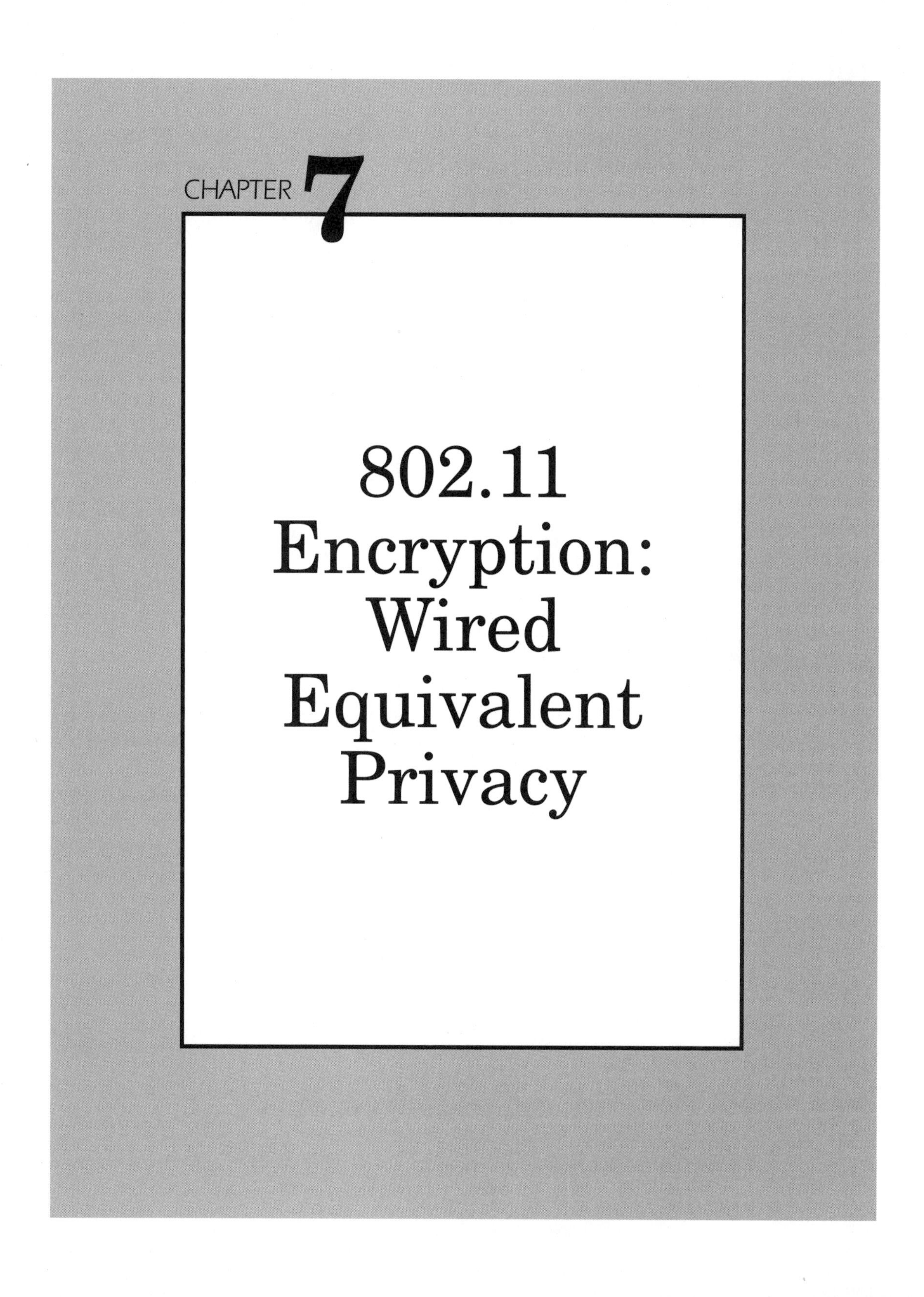

CHAPTER 7

802.11 Encryption: Wired Equivalent Privacy

How does one effectively deploy a wireless LAN to ensure proper security measures have been taken? The answer lies in deploying all points along your network so that you maintain the same consistent type of security as you would with wired LANs or dial-up connections. This leads to the concept of WEP.

WEP is an acronym for wired equivalent privacy, a concept developed as part of the IEE 802.11 standard. WEP offers the same level of privacy that you would expect to maintain in your wired network. The 802.3 Ethernet standard offers security protections for a wired network through physical security means. Since you are only dealing with wires per se, you can control who has access to your network room by simple lock and key.

Because you can physically exclude outsiders from a wired network, the wired LAN standards need not necessarily offer encryption to protect your data against someone interested in trying to view your network data traffic. But because wireless LANs are not protected by a physical space, any transmissions can leak beyond your office building and literally right out into the street.

Why WEP?

You wireless LAN can defend against most forms of eavesdropping, but the only way to prevent any hacker from compromising the integrity of your transmitted information is to use encryption, and that is where WEP comes in. WEP makes certain that most WLAN systems have a sufficiently high level of privacy (that is analogous to that of wired networks) by encrypting the radio transmissions. WEP also prevents any unauthorized users from accessing your wireless network through the means of *strong authentication*, which is not normally a part of the 802.11 standard but crucial to using WEP.

WEP also provides for access control through authentication, meaning that most 802.11 wireless LAN products support WEP as one of their core set of features.

Defending Your Systems

WEP is your method of defending your systems from eager eyes trying to view your important data. The best procedure is to access the settings for your wireless 802.11 network and make certain that the first thing you do is turn WEP on. Most users are simply not aware that encryption mechanisms are already built into their networks and as a result fail to take even the easiest precautions to make sure that their data is encrypted.

There are several methods by which you can change and manage your WEP key. Remember to change the default encryption key that is in your router or wireless LAN. Change this key often because, given enough time, an eager hacker can break your encryption key and still be able to view and access your 802.11 network.

Should someone gain access to your system, you can take very easy steps to ensure that your internal network data assets are protected as well. You should always password-protect your hard drives, network folders, and any other assets on your network so that you make it that much harder for someone to view or access your protected data.

Every wireless station has a wireless network name called an SSID. You should take the very easy step of making certain you change the default name immediately. Most 802.11 routers are preconfigured with a standard encryption key and SSID to get you up and running quickly. It is a simple matter for a hacker to know the settings your 802.11 router and quickly configure his laptop with the same default settings to access your network. In fact, he could theoretically take a laptop, sit just outside your office building in a car, and gain full access to your wireless network, and you would never even know about it.

If your 802.11 network allows for the use of *session keys*, you should take advantage of them because they are just another step to ensure that each network session is encrypted.

One easy way to keep a sort of physical control on your network is to use MAC address filtering, if you have that option. In this way, your router will not accept network connections from any computer that you have not already specified in advance. Every network card has a unique MAC address, much like a social security number is unique to an individual. You can easily enter this MAC address into the router, so that any other computer that has not been cleared for access cannot access your network.

VPN systems are an excellent way of making certain you have a virtual private and secure network connection within your wireless infrastructure. VPNs offer greater security and keep a direct connection between the client and the host computer. However, this often requires a specialized VPN server. On a more positive note, most Windows operating systems (including Windows 98 SE, Windows 2000, and Windows XP) already have a built-in VPN client, making it that much easier for you to roll out a VPN.

Data is the lifeblood of many organizations, and you most likely require a very high level of protection to keep your data secure. To employ extra security measures, there are methods that involve Kerberos and Peer-2-Peer encryption mechanisms. Using the following methods will assist you in making certain you have taken at least the most basic measures to ensure you are protected, as shown in Figure 7.1:

1. Encryption from point to point
2. Strong password protection
3. User authentication
4. Virtual private network (VPN)
5. Secure socket layer (SSL)
6. Firewalls
7. Public key infrastruction (PKI)

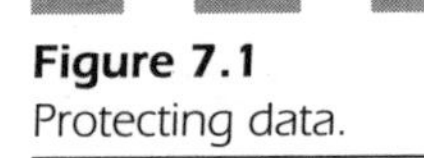

Figure 7.1
Protecting data.

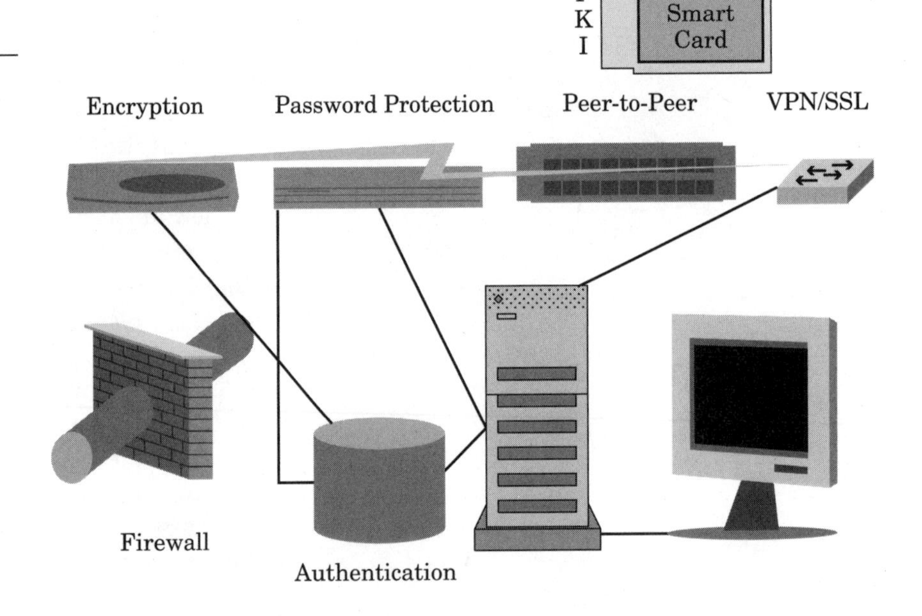

Future directions for 802.11 involve extending WEP to integrate future standards developed by the IEE 802.11 Task Group. These enhancements will more than likely involve new and more secure mechanisms, thus making it possible to deal with new threats that are constantly evolving in this insecure world.

WEP Mechanics

WEP is designed to prevent someone from casually eavesdropping or modifying any portion of your data stream. WEP uses an RC4 40-bit stream cipher to encrypt data and a 32-bit CRC to verify it. Unfortunately, it has a faulty algorithm, so that several types of attacks can succeed against it. The biggest problem with the algorithm is that RC4 is subject to key-steam reuse, which basically destroys the ability for it to encrypt information effectively. Attacks against the RC4 algorithm involve collecting frames for statistical (traffic) analysis, using SPAN to decrypt frames, and "flipping" data so that messages and information are altered.

Wireless Security Encryption

WEP uses a secret key shared between the wireless user and the access point so the all data transmitted and received between the wireless station and the access point may be encrypted using this same shared key. 802.11 permits the use of an established secret key unique to each wireless user. In the majority of cases, one key is shared throughout all users and access points on the WLAN.

Data encryption is defined using weak (40-bit) or strong (128-bit) classifications, as shown in Figure 7.2.

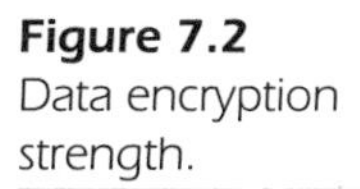

Figure 7.2 Data encryption strength.

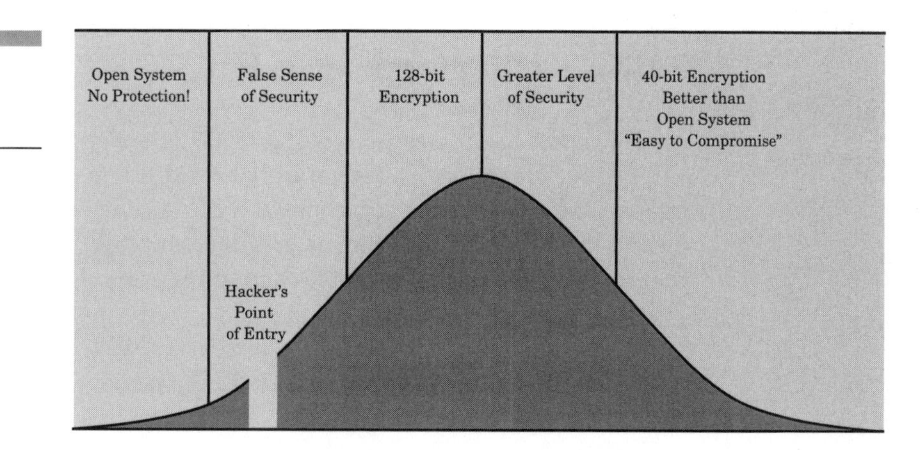

Encryption is comprised of the secret key and an RC4 pseudorandom number generator. Normal data is first encrypted and then protected against unauthorized modification while it moves across the network.

The secret key is transformed into the final shared key that is inserted into the pseudorandom number generator (PRNG).

Insecure Keys

The insecurity with keys is that they are more often than not shared across all stations and access points in the network, so that key distribution is a major problem. Note that when you take the same key and share it with a number of users, ultimately that key will not stay secret.

Key insecurity is addressed by configuring the wireless stations with the secret key, as opposed to allowing the users to execute this procedure. This is still not the best answer, because the shared key is stored on the user's computer where a hacker can potentially retrieve it and use that key to access the network fraudulently. If this happens, then all the keys saved on every other wireless user's computer must be reset with an entirely new key.

The best way to defend against insecure keys is to migrate to a system setting that assigns a unique key for each user's computer; you should still change the keys frequently, because you never know when that key can become compromised and lead to an open avenue of opportunity for a hacker to gain access to your wireless network.

Taking a Performance Hit

As you might have guessed, adding levels of encryption to your network will ultimately reduce your overall bandwidth and slow down your wireless connection speed. Even though WEP is considered a fairly efficient means of adding encryption, it is important to quantify exactly what that level of security is going to cost you in terms of speed.

- 40-bit encryption reduces bandwidth by at least 1 Mbps
- 128-bit encryption reduces bandwidth by nearly 2 Mbps

Even at full signal strength with speeds of 11 Mbps, you will notice the drop in speed whenever you start transferring files, sending large documents to your networked printer, or storing any large document on a file server. In addition, when you are not at full signal strength and you have already reduced speed throughput because of your distance from the wireless router, the performance hit becomes more obvious in terms of how long it takes to transmit multimedia files or browse graphically intense Web sites over the Internet.

Wireless Authentication

The two levels of WEP authentication (Figure 7.3) are:

- **Open system**—This scheme allows all users to access the wireless network.
- **Shared key authentication**—This is the more secure mode that controls access to the wireless LAN and stops hackers from reaching the network.

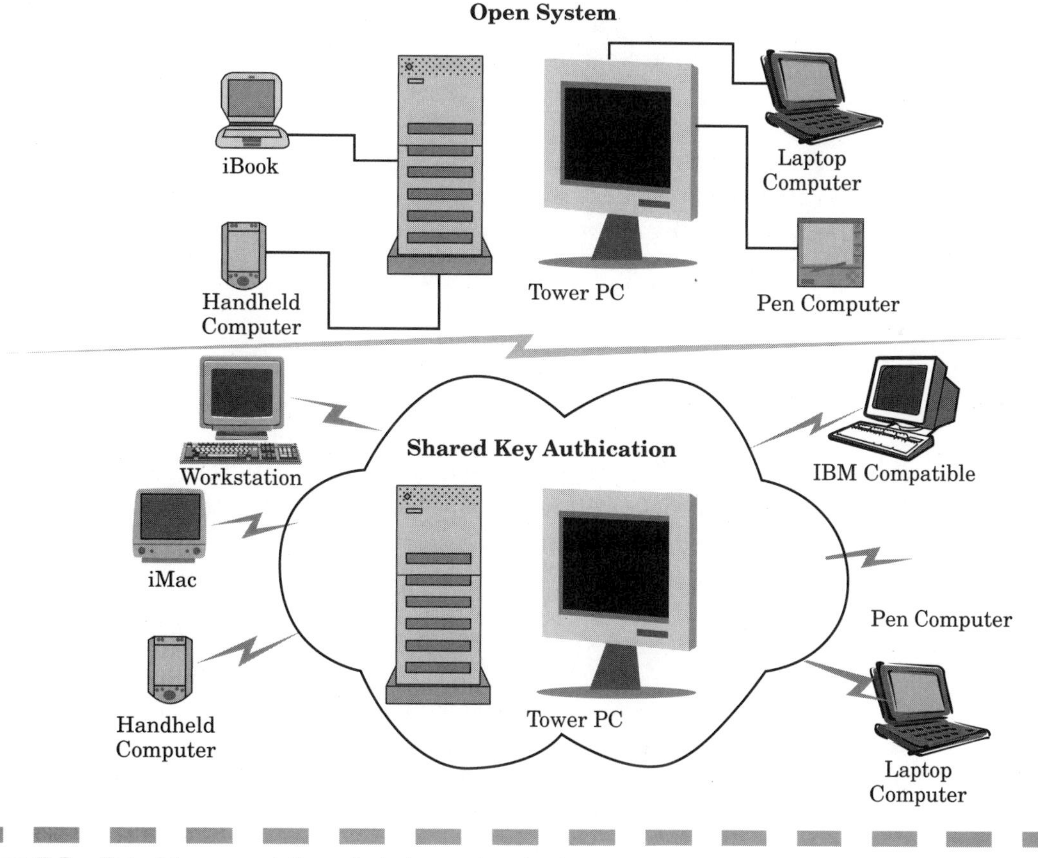

Figure 7.3 Pictorial representation of wireless authentication.

Shared key authentication uses a secret key that is shared throughout all wireless network users and access points. Whenever a user attempts to connect to an access point, it will reply with a random text to challenge the user's machine to identify itself as being authorized. The wireless workstation must use its shared secret key to encrypt this challenge text and reply to the access point in order to authenticate itself to the WLAN. Then the access point will decode that response using the same shared key and compare it to the challenge text it used before. Only if the two results are the same will the access point confirm that the wireless user can log into the network. If, however, the wireless user does not have the same key or responds incorrectly, the access point will reject any access attempt and prevent the remote user from accessing the network.

It is important to know that WEP encryption is possible only in tandem with shared-key authentication. However, if these precautions are not enabled (and they are not by default) the system will function in "open system" mode that allows anyone within in range of the access point to gain access. In these very circumstances, hackers prey upon the weaknesses of your wireless system.

Everyone on your wireless network may use the same shared key, but even with this authentication enabled, authorizing just one individual is not possible because everyone is considered one group using the same shared key for network access. If you have several users in your organization, then this "community key" can be easily acquired and there is a greater chance for an unauthorized user to access your network resources.

In most cases, the key used to authenticate users is the same as that used for encrypting the data. This can constitute a major security breach for any wireless user, regardless of platform. When a hacker has a copy of the "shared key" he can use it to access your network and view other users' network traffic. This causes even greater network problems.

The best defense against this type of problem is to send out separate keys to be used for authentication and encryption in your system (Figure 7.4).

When you keep these two keys separate, you increase your chances that a hacker will not be able to compromise the mission-critical data traveling across your network even if he does gain access and log onto your system. In short, you can never be too secure. Don't reuse the same keys for the sake of convenience, because this compromises your security.

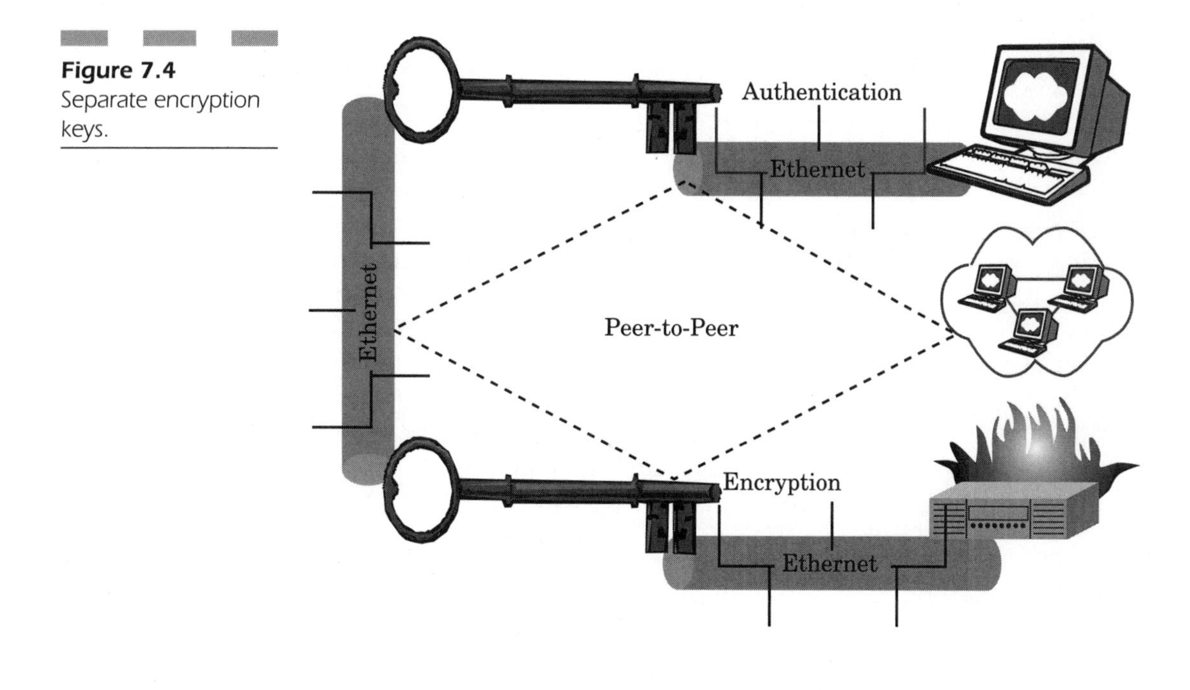

Figure 7.4 Separate encryption keys.

Known WEP Imperfections

There is a major problem with WEP: it has a number of imperfections that make it highly detrimental to any serious security concerns about protecting your WLAN. Due to this fact, WLANs using WEP are susceptible to being attacked in a number of new ways.

WEP also suffers from being vulnerable to not accounting for unauthorized traffic or decryption that may result from a hacker who is trying to log in fraudulently to the access point in your WLAN. These problems make WEP a poor choice as the only means of protecting your network against possible intrusions. In order to defend yourself appropriately, you should maintain a virtual private network in combination with using WEP, so that when a hacker "cracks" the encryption scheme for WEP, he would also have to break the encryption scheme of the VPN carrying the individual packets of network traffic. If you make it much more difficult for the hacker to do his job, you have a reasonably greater measure of security protecting your WLAN.

Access Control

It is important to consider all types of access control techniques that prevent unauthorized use of your wireless network by hackers. In addition to the shared-key authentication method mentioned in the previous section, there is another technique, *extended service set identification* (ESSID). This represents an alphanumeric value programmed into a wireless router to determine which subnet on your wired LAN it is part of. This value is used as a means of authentication to make certain that only authorized wireless users can access the network. If the wireless user does not know the ESSID, he cannot use the network.

However, most wireless users can tell their network interface card to enter "promiscuous mode" in an attempt to try and automatically determine the ESSID. This is easily accomplished by setting the parameters for the ESSID on the wireless computer without any value (null) whatsoever. In this way, the wireless card will enter promiscuous mode and automatically roam until it finds a wireless network to access. This is the method by which most hackers gain access to computing systems.

Another means of controlling access is to tell the wireless router to screen out any wireless network interface card that does not have a particular media access control (MAC) or machine address. This is a very good form of access control that will prevent unauthorized users who set their cards into promiscuous mode from entering your network without prior authorization.

These MAC addresses are retained on an access control list (ACL) that is part of the wireless router or access point's configuration. The parameters are usually set by the internal Web server within these devices. The router examines each unique MAC address and only allows authorized MAC addresses to log onto the network. This form of control effectively limits the access to your network to those stations that are authorized; anyone else is rejected.

Administrators can enable this extra form of security to exclude hackers from outside wireless computers as well as those users who are part of a different network within your organization. By segmenting users into pools, you can restrict access to wireless servers to those people who have a "need" for access.

Note that it is possible to "spoof" a MAC address so that a hacker's wireless computer appears to be an authorized machine logging onto your network. This is why it is important to maintain a log of all traffic coming in through your wireless network, so that you can determine if there are spikes in activity that don't belong. Armed with this information, you can keep a watchful eye on your network for unauthorized hacking activity and protect your mission-critical data.

IRL Security

IRL refers to "in real life" security; while most hacker attacks occur on the "ether" of the net, IRL security refers to physical security dealing with the actual nuts and bolts of your system. The concept of protecting these resources is as real as any other element of your system security.

The tangible portions of your physical network can be damaged in any number of ways, not the least of which is sabotage. Wireless networks have all the same types of problems that any radio station endures. Lightning strikes can disrupt your equipment and cause irreparable damage.

Access points are essentially radio transmitters, so it is important to ground all your equipment and locate the antennas in areas that are not near the outside areas of your building.

Hackers may see exposed antenna arrays as an easy way to reverse transmit radio signals through the antennas in an attempt to destroy the wireless transmitter in your equipment. In fact, most radio transmitter assemblies can be tweaked to transmit 100 watts of power into a transmitting antenna. The purpose of this action is to destroy your transmitter and your WLAN. This is easy to do if any portion of your radiating assembly is in an unsecured area that a hacker can access.

Points of Vulnerability

Hackers know they want to corrupt your wireless network, and it doesn't take too long once they gain physical access to your corporate offices. It is common that a cleaning staff person can actually be a hacker who needs only a few minutes alone in your network room to destroy cabling, install a virus onto the server, fray the cables (causing intermittent connectivity), or simply pour water onto your computing devices.

The best defense against physical vulnerabilities is to keep access to any room with WLAN equipment away from everyone except authorized personnel who are supposed to be working in this equipment. You should always place access points and associated antennae in a secured area that is nowhere near public places. You should then protect this equipment by placing the machines in locked rooms with barriers and access controls that prevent anyone from getting near this equipment. You may also wish to install intrusion detection systems (IDSs) that monitor activity near this equipment and watch these assets with remote cameras used by special administrative personnel over the Internet (through a secure channel!) or at a remote office location so that you know who has been accessing your wireless equipment in an attempt to gain unauthorized access.

Keeping track Any person with a wireless network interface card constitutes a potential risk of allowing a hacker to gain possession of an important key and use it against you. This is why it is vital to keep track of all your wireless users by using administrative types of controls to log and record users who have wireless computing devices. You can compare this information against the log information from your access point to track down any fraudulent activity.

Wireless policy What kinds of measures can you enforce for wireless users to maintain a higher level of security? You should never allow users to leave their computers alone in a public place where anyone can use the machine or alter its contents.

Always establish a wireless computing policy. Computer theft is very common and a hacker can easily use a few minutes alone with a machine to gather all the information from it that he needs to institute a viable attack against your WLAN.

The policy you create can be a written policy that outlines all the procedures and methods important to apply when securing your laptop or other wireless PDA.

Simple precautions would require that the wireless user log out of the WLAN so that someone who did try to access your WLAN would not be able to log in immediately without the proper authentication information from the user himself.

In addition, your policy should not permit users to take their wireless devices with them to a public place where they eat lunch. Just getting up and finding a soda can leave your machine alone long enough for someone to compromise its link to your WLAN. If you do see someone who appears to be looking at your wireless device or trying to copy down information, activity such as this should be reported to the administrator right away.

You can never be too careful with the security (or lack thereof) of wireless devices!

Conclusion: Finding Security in an Unsecured World

This chapter has focused on the concept of how a wireless network can effectively provide the same level of security as that of a wired network. However, the most important section of this chapter details all the ways in which the wired equivalent privacy (WEP) can be circumvented through several methods (Figure 7.5) including:

- IRL in real-life physical damage
- Jamming and interference
- Eavesdropping
- Unauthorized access

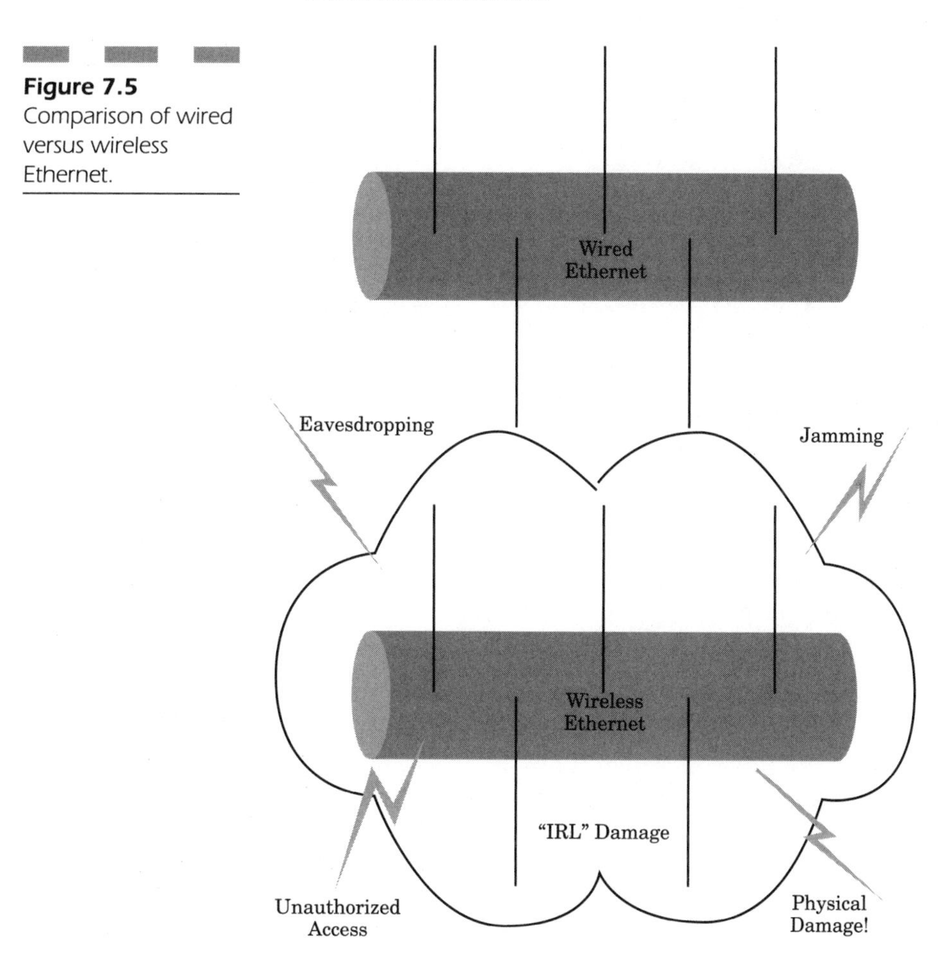

Figure 7.5 Comparison of wired versus wireless Ethernet.

By understanding these methods of attack, you can determine the best manner in which you can prevent a hacker from accessing your system and compromising the WEP of your wireless network. Using WEP encryption enables your WLAN to resist the way in which encryption keys are stolen. Effectively managing these keys maintains the wired equivalent privacy of your WLAN so that you can use your network with the confidence of maintaining a level of security analogous to that offered by a wired LAN.

Physical barriers and protection mechanisms are the most important means of making certain that hackers are unable to gain access to precious network resources.

What equipment is vulnerable?

- Access points/wireless routers
- Laptop computers
- Wireless PDAs
- Wireless network interface cards
- Network printers
- Network file servers
- Network fax servers

Effectively maintaining physical safeguards prevents hackers from getting into these systems in person, and therefore makes it harder for them to log in wirelessly. The idea is to protect wireless users as well as the actual access points; only then can you be certain to manage all your network resources effectively.

Finally, maintain a log of every activity on your WLAN. Keeping this information protects you from hacking activity. Your best defense in keeping your network secure is to make certain you can identify malicious activity on your system. If you maintain a vigilant eye on your resources, you can effectively protect your wireless network against many possible intrusion attempts.

Following these guidelines will enable you to operate a secure wireless network without fear of being hacked.

CHAPTER 8

Unauthorized Access and Privacy

The 802.11 standard uses the wired equivalent privacy (WEP) protocol. This protocol is intended to provide authentication (prevent unauthorized access) and privacy (prevent data compromise and data tampering) equivalent to that in a wired connection, but the 40-bit key is too short to prevent data compromise. Additional points of concern about WEP and the standard are:

- The 24-bit WEP initialization vector (IV) is too small to prevent repeated use of a cipher stream.
- The manner in which the IV is used is not specified in the standard.
- The integrity check value (ICV) is useless for detecting alteration of frames.
- SSID data is not protected by encryption.

Privacy in Jeopardy

Privacy is an important element that ensures that information does not fall into the hands of any unauthorized computer user. However, one of the biggest problems with Wi-Fi is that radio waves carry confidential data and by their very nature can be intercepted. Unlike wired Ethernetworks, Wi-Fi is accessible at a distance with the proper eavesdropping equipment.

Passive Attacks

Hacker exploits use a "passive attack" (Figure 8.1) to intercept the signal from an 802.11b network in order to acquire information that includes:

- Network IDs
- Passwords
- Configuration data
- Mission-critical or confidential user information

The reason that risk is so prevalent in 802.11b is that this Wi-Fi variant transmits signals that can easily penetrate building materials because it uses the 2.4-GHz radio frequency spectrum. The 802.11b transmissions can be easily detected from outside the building, away from plain sight.

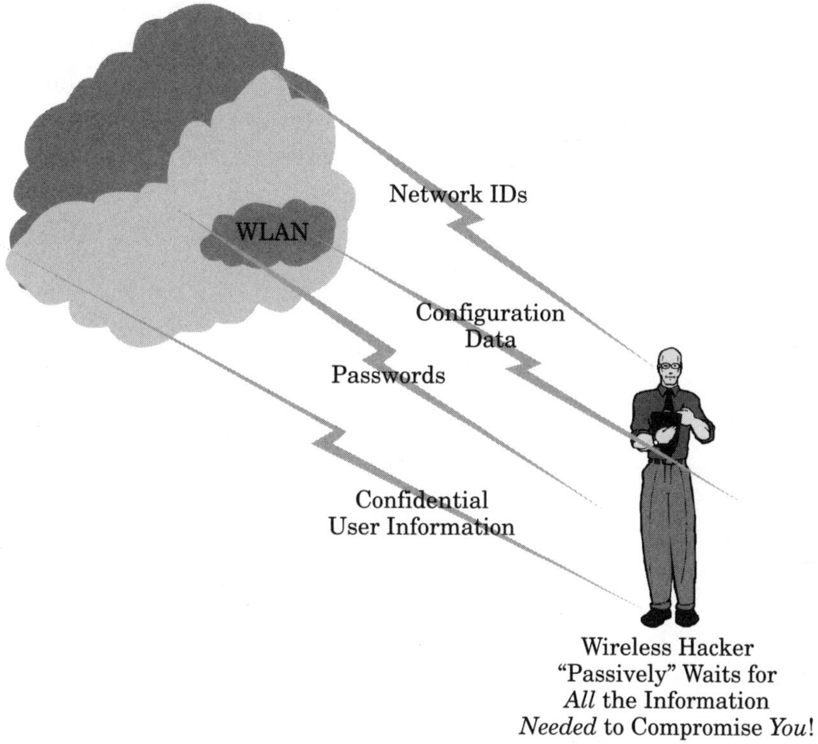

Figure 8.1
Passive hacker "waits" patiently for your information.

This type of attack is executed using a "sniffer," a wireless network analysis tool. The ease of this attack is well known because most 802.11b networks don't even use the most basic security measures and because there are quite a few vulnerabilities within the weak encryption protocol that hackers can easily break, given enough time.

Broadcast Monitoring

Privacy is easily compromised through "broadcast monitoring," a form of eavesdropping. Hackers can set a mobile computer to search out any wireless network and monitor all broadcast traffic.

When an access point is part of a hub, it is well known that the hub normally broadcasts all network traffic to all connected devices. This makes all network traffic vulnerable to hackers who wish to monitor your data channel. This means that a wireless workstation can monitor broadcast traffic that could be meant for literally any other client on the wireless network so long as the access point is directly attached to an Ethernet hub. In order to reduce this problem, it is strongly recommended that you use a "switch" as opposed to a "hub" when connecting your access point. A switch dedicates all packets meant for only *one* wireless device during any transmission. However, a hub sends out all transmission packets at once. A switch is a dedicated form of sending traffic, while a hub allows all connected devices to "share" the bandwidth, leaving your privacy in jeopardy from anyone who realizes this vulnerability.

Active Attacks

Wi-Fi privacy is immediately compromised from an "active attack" whereby a hacker uses a program that "sniffs" the airwaves of a wireless network to acquire confidential information (Figure 8.2) such as:

- User names
- Passwords
- Any personal data

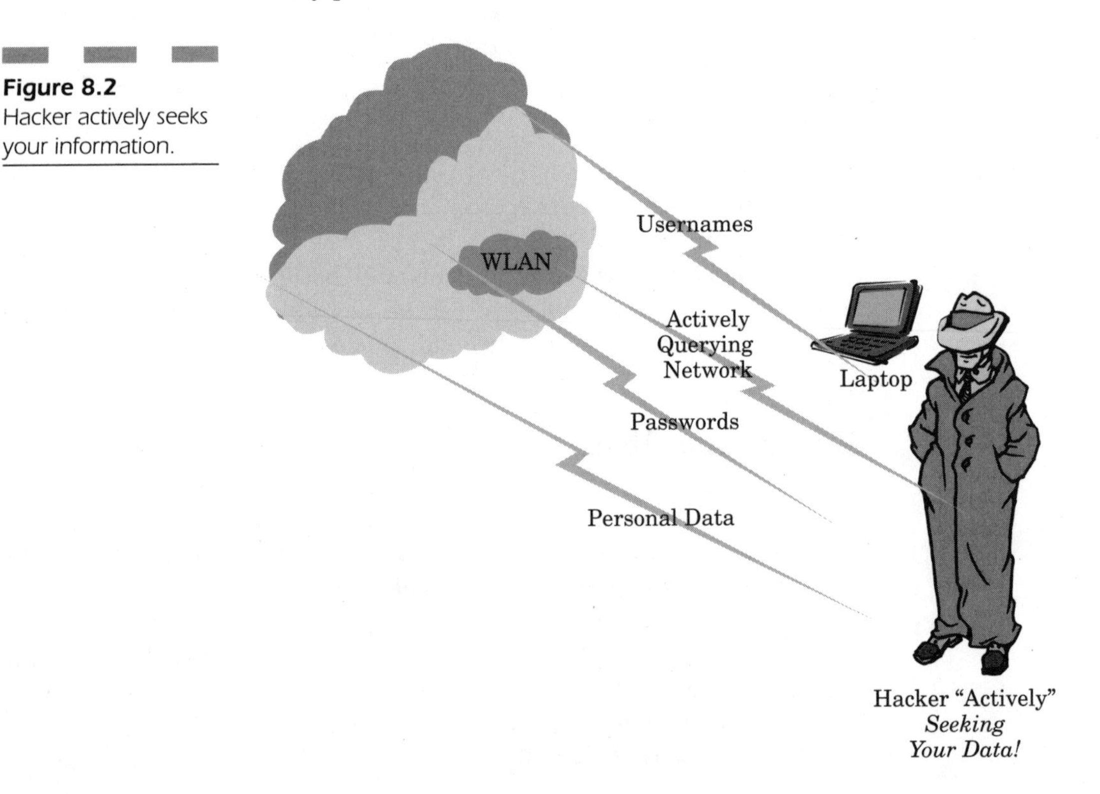

Figure 8.2
Hacker actively seeks your information.

In this type of attack, the hacker can use the information acquired to pretend to log onto the network as an authorized user with the ability to access any wired network resource on your intranet. Once the hacker accesses your network, he can map out your entire internal infrastructure using software that is freely available from a number of hacker sites. This leaves your mission-critical information completely vulnerable to theft, modification, or worse, being erased from your file servers.

The "Evil" Access Point

One of the sneakier tricks that hackers use is to plant an "extra" access point somewhere in the corporate facilities (or close to a pocket of heavy wireless network traffic) in an attempt to capture wireless traffic without the knowledge of the wireless user.

This attack is difficult to discover since the access point is hidden from view. The only requirement is that the "evil" access point have a stronger signal than the "true" access point. This device could easily capture enough information, which the hacker can exploit as an access vulnerability, to access all the wireless network resources without proper authorization mechanisms.

Data Privacy

Wi-Fi networks have a number of issues when it comes to dealing with data privacy. Wired Ethernetworks do not have as many data privacy issues due to the inherently secure nature of the lines that carry data. You must be physically in contact with the wired network to acquire information, which is not the case in the Wi-Fi world.

Privacy becomes a serious issue when an unauthorized user gains access to something as simple as text e-mail. That person could use unauthorized access either to erase or to alter the data in a message just by connecting through the wireless network. This action would leave not one, but all e-mail suspect; corporate information would lose any trust relationship with customers, employees, and executive staff. Since e-mail is a cornerstone of doing business today, no information would be deemed safe for any business task.

There are a number of "active attacks" described earlier in this chapter that demonstrate how the security functionality within the 802.11 standard does not offer a sufficiently secure means of maintaining data confidentiality. The primary reason for this type of problem is that the WEP means of encryption is only a straightforward linear cyclic redundancy check (CRC) mechanism which can be easily fooled under an attack where the message and cryptography checking methods are altered. Authentication records and hash mechanisms can be fooled, making the user think the message was transmitted without being breached, when in reality the entire text of the message could have been changed.

Compromising Privacy in Public Places

Wired networks literally know no bounds. Other companies may have wireless networks across the hall or on the floor right above you. The frequencies used by Wi-Fi networks easily penetrate building materials to the point where someone could easily use your wireless network from a nearby location not in your internal corporate environment.

However, what if you have a legitimate business need to access your internal corporate network from a public environment outside your corporate offices? What if you are a mobile employee, are working at a remote client site, or simply have the need to work out of your home office one day? All these places are more accessible to the public, so how does that affect your security?

It is a business necessity today to access network resources remotely, and at times that does involve using wireless links, while at other times you need to use external networks or ISPs that are beyond your scope of trust.

For example, many airports and corporate networking centers allow mobile road warriors to use their existing laptop equipment to connect (using third-party wireless networks) at a fee, as a conduit into their corporate intranet. Today, you will find this service more commonly at Internet cafés, networking conference centers, and in many large airports. More and more organizations are deploying a simple 802.11 infrastructure to support mobile customers.

Protecting Your Privacy

One fundamental way to protect your traffic over unsecured networks is to use a corporate virtual private network (VPN) to enhance your security through public network facilities. With this type of setup, if anyone were to attempt to monitor your communications over your wireless link, they would not be able to make sense of your traffic because it is encrypted using a special sequence unique between your wireless workstation and the server at your company's headquarters. This type of encryption is independent of what you find using the WEP encryption that is part of the 802.11 standard.

The risks associated with using a third-party network (Figure 8.3) include the following:

1. Public wireless networks use a strong power level when transmitting. This can become a serious vulnerability when hackers can eavesdrop on your data channel from almost anywhere in the near vicinity.
2. Public networks, by definition, can be accessed by anybody. Any traffic you send can be monitored, altered, or disrupted in any number of ways. You have no control whatsoever over your network connection.
3. Public networks function as "links" to your network. This means that if a hacker were to gain access or acquire your network passwords, he or she could then potentially access your internal network through a spoofed connection. This would leave your entire corporate infrastructure vulnerable to attack in the worst possible ways.

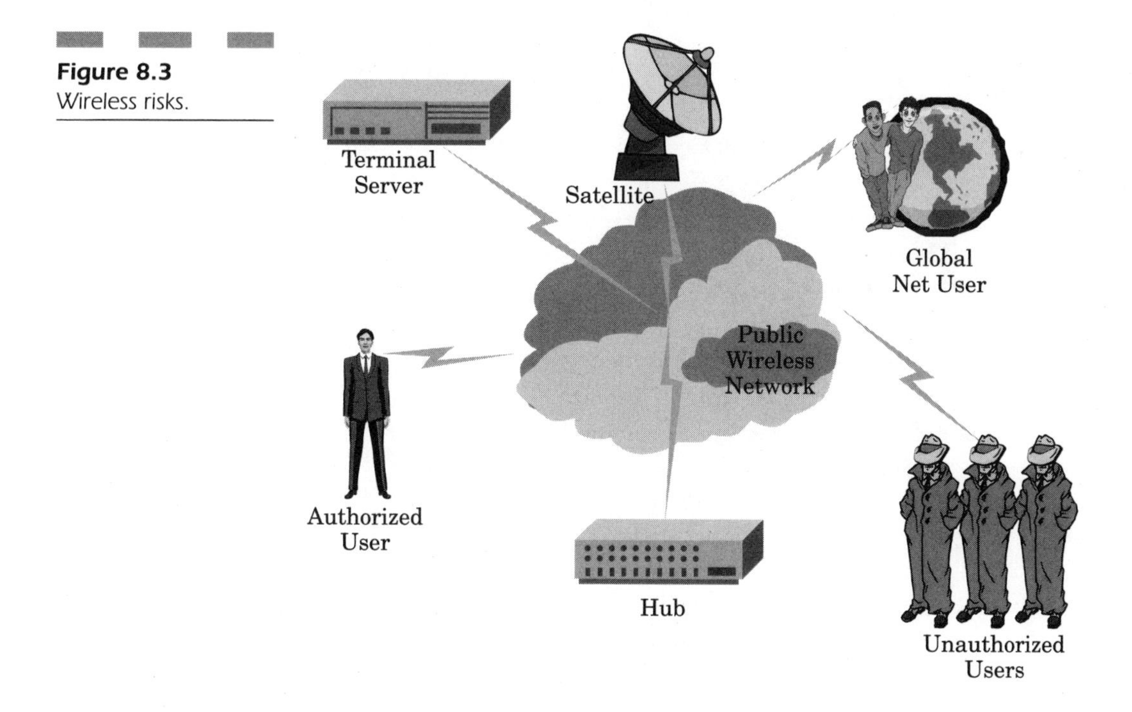

Figure 8.3
Wireless risks.

When using a third-party network, you expose potential access gateways into your fixed Ethernetworks too. Your only defense is to be wary of these vulnerabilities and take the necessary precautions to monitor incoming connection traffic from any external links into your server.

Public or Private?

Many organizations successfully reduce their risk to external connection by designating specific resources as either public or private. Private resources are NOT accessible through any external connection, and if all else fails with respect to security, those protected resources will ensure that you do not go out of business from being hacked.

Public types of resources can be protected by using an application-layer security protocol. For example, transport layer security (TLS) is a flavor of the more commonly used secure sockets layer (SSL) that you use whenever performing a purchase transaction over the Internet so that your credit card information is not intercepted while in transit from your workstation to the server.

Private resources are protected either by permitting only local wired connections to access them, or by creating the next best thing—VPNs that make certain that even if those resources are intercepted they are unusable to the hacker because they are encrypted in a strong and secure manner. VPNs are one means of making certain that wireless hackers are unable to eavesdrop or access unauthorized private resources anywhere on your network.

Safer Computing

Inasmuch as this chapter has focused on the ways in which hackers can compromise your wireless network, the one element that will save your corporate infrastructure is understanding specifically how to counteract these threats and shore up your vulnerabilities. There really isn't any such thing as 100 percent security, but you can take certain precautions to facilitate a safe computing environment.

Security often does not come cheap (Figure 8.4). In fact, most security measures are never even implemented because they are a function of money. Some security measures are not used because they are too expensive, but the real risk is in the expense of being hacked. Common reasons not to add security measures include:

- Expense in added security mechanism
- Extra maintenance
- Higher learning curve
- Inconvenience
- Extra operating expenses
- Financial risk

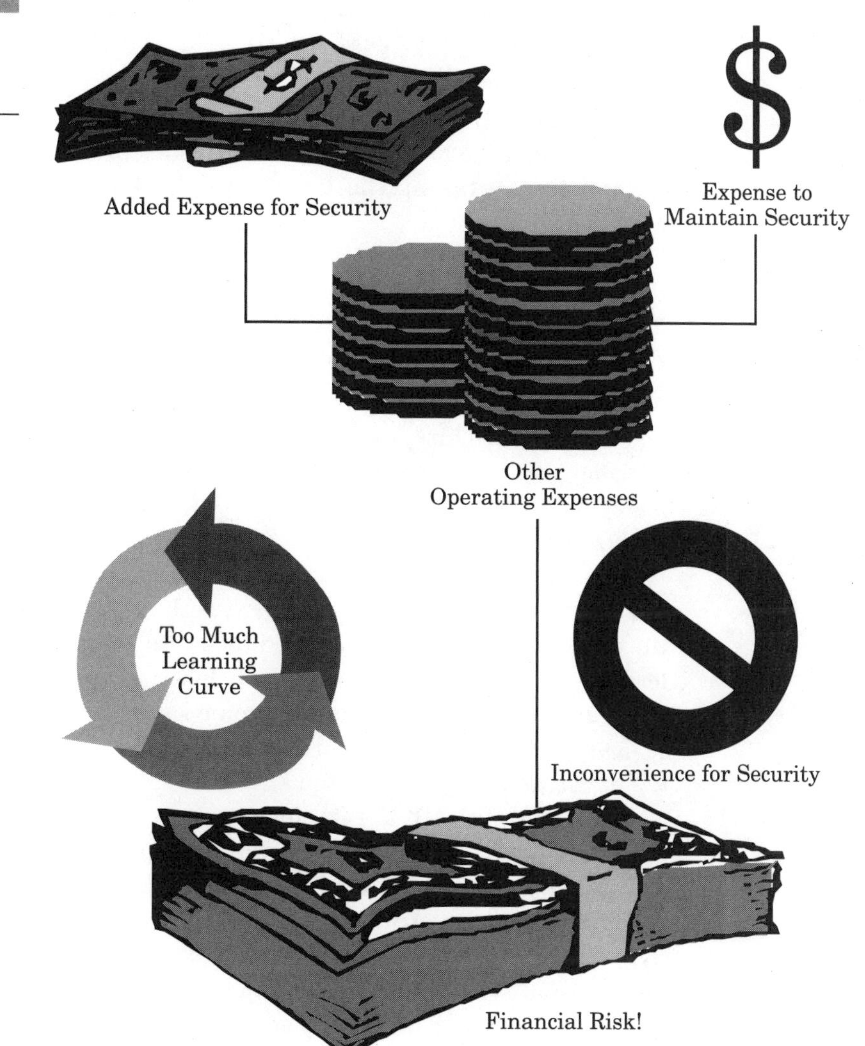

Figure 8.4 Why companies don't have security.

The "Human" Factor

Security is not simply a measure of adding expensive equipment and more machinery, and educating employees to watch out for security at the expense of the position they were hired to do. The most effective security often involves the "human" side of the equation, where employees can follow certain guidelines in a well-defined security policy. Taking simple precautions and steps as outlined in a corporate policy for using the WLAN is often very useful in actually preventing hacker attacks, and these steps do not detract from the employee's time.

Security policies are often a comprehensive list of all the preventive measures you can take with respect to two main ideals: technical expertise and operational procedures.

The biggest question you may face is getting executive management on board with the idea, signing onto the idea of "needing" a security policy, and finally taking the time to implement the guidelines with all wireless users.

Defining the Bullet Points in a Security Policy

Some companies hire a special security consultant to write their security policy, while others pay close attention to their needs and assemble a policy based on the "best practices" they have learned when connecting to their wireless network.

When you create a security policy, you must first determine who has a legitimate need to use your WLAN, both internally and externally. For those people who have a real need to access your network resource wirelessly, you must determine if actual Internet access is required. If it is not required, then you are adding some safety precautions by not having to deal with external Internet connectivity that acts as a portal for hackers to use to access your systems.

When dealing with the implementation issues within your company, you must pay close attention to defining specifically who has control of any access point installed in your organization. It is then very important to define specific restrictions for placing any equipment in your company. Determine who can potentially access the physical location of any piece of your wireless equipment. It is best to hide all such equipment so that it is far less possible for someone to alter the settings to permit unauthorized access.

Your security policy can also function to protect you by defining what type of information is permissible to send over your WLAN. It is important to designate the specific types of wireless devices that can connect to your network.

When working with access points, make certain you have clearly defined security functionality. Password protect your internal configuration Web pages so that nobody else can modify your security settings to detract from them. You can then proceed by defining exactly what restrictions you place on each type of wireless device. This means you can specify certain locations where mobile devices may or may not be used. The idea is to keep as much control as possible so that you know who has access to each device on your WLAN. This information helps you keep track of your network access so that you can more easily track down potential hackers attempting to breach your network resources.

Policy guidelines When writing your security policy, it is very important to provide as much detail as possible. The guideline is simple, "be as specific as possible" and try not to leave any room for interpretation. A security policy is designed to be a method of protection.

For example, when you describe your hardware and software configuration, include as much detail as you possibly can about each mobile device that will access your WLAN. You should include the device configuration, unique wireless MAC address, and specific login credentials that let you know exactly what type of device you are communicating with. The idea is to maintain as much knowledge as possible about the devices on your network. If someone tries to spoof a device, you have a reference point that more quickly allows you to determine discrepancies in the connection that would indicate a hacker trying to breach your network safeguards with an unauthorized device.

You can also use your security policy to immediately block out the connection parameters from a mobile device when it has been stolen or its login information has been compromised. Employees need to have a clearly defined procedure that allows them to report the loss of any wireless workstation or PDA as well as any security breach where its information may have been compromised.

The security policy also dictates connection safeguards that involve the use of encryption as well as other security safeguard software meant to protect your network against possible breaches.

Timing is also an important element to specify within your policy guidelines with respect to how often and comprehensively your organization will perform a security vulnerability assessment. It is important to understand that there is no such thing as 100 percent security, and the fact that devices, drivers, and software change all the time contributes to the weakening in your security. New vulnerabilities are found almost every day in computer operating systems, hardware, and network connectivity. It is very important that you have a schedule of ongoing security vulnerability assessments and continually scan and monitor your computer systems for ways in which hackers can compromise your network. Information is power, and that power translates into your ability to plug any security holes before a hacker finds them and uses them against you.

Training

Inasmuch as you can install the most expensive security software, write up the most detailed security policies, and implement the strongest level of encryption to protect your wireless traffic against eavesdropping—the most important element of maintaining security will always rest on how well your employees are trained to deal with potential security breaches.

It is of the utmost importance to make certain that all of your mission-critical personnel are correctly trained on how to use wireless networking protocols in the most effective and secure manner possible.

For example, network admins needs to go through a comprehensive security training course so that they understand and realize the needs and changes of their WLAN. In order to acquire this knowledge, they must comply with security policy and understand the exact and proper procedure to take when they realize that a hacker attack is "in progress." Since the most effective means of protection is a trained person with an acute awareness of security, it is also very important that these staff members have continuing education so that they can deal with the constantly changing world of security and know how to deal with new threats as they arise.

Physical Security

When it came to security for your wired network, physical security was the most important part of making certain that hackers could not pose as legitimate employees, enter your corporate facilities, and use "sniffer" tools to acquire data from the actual wires in your network. With the advent of wireless networks, physical security was not considered as critical as it was when dealing only with wired networks, because the person did not need to enter your facilities to attempt to breach your access safeguards.

The truth is that physical security is just as important with WLANs as it is with wired LANs. This level of security is the most basic step of making certain wired or wireless equipment is configured so that it permits incoming wireless access only from authorized users who are meant to use the network.

Physical security measures are vital because they contain important methods making certain you have the necessary:

- Identification protocols
- Intrusion detection
- Access controls
- Logging facilities (so you know who accesses your network and when)

Even wireless networks must support physical access controls so that not just anyone can waltz into your facility and start reconfiguring your devices. Identification devices include:

- Card readers
- Badges
- Photo IDs
- Biometric identity devices

Access methods In order to control more elements of your physical access, it is highly advantageous to add as many locked areas as possible. Most companies lock up their server rooms, but how many access points are there to workstations within a normal company?

Your external facilities should have barriers around their perimeters including locked doors and video surveillance cameras.

These methods function as active deterrents to any hacker who might be interested in attempting to access your wireless network from outside your building. The hacker would think twice about lingering around your parking lot if he believes there is a video camera taping movement for possible future prosecution.

This is why biometric devices are advantageous (Figure 8.5). You can configure laptops (or any wireless device for that matter) not to turn on unless you have properly authenticated yourself to the device via a biometric access mechanism. These access barriers include:

- Scanning the iris or retina of your eye
- Determining the geometry of your hand
- Scanning your palm for unique features
- Scanning your fingerprints
- Checking your unique voice pattern
- Face recognition (a growing field since 9/11)

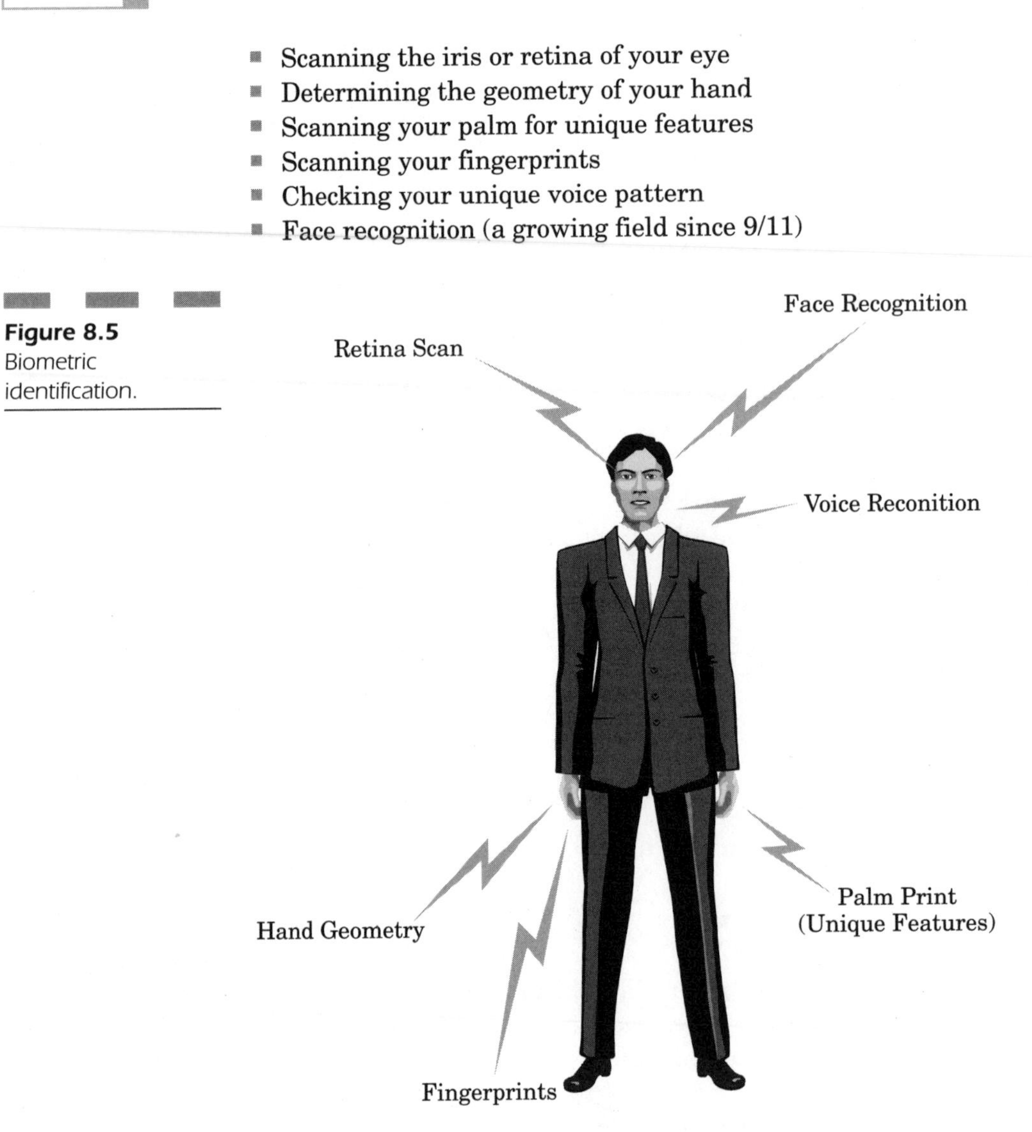

Figure 8.5 Biometric identification.

Wireless Range

Unauthorized access is most commonly obtained when you have multiple long-range access points installed throughout your company. These access points function to increase range so that wireless users can always access network resources, but the negative aspect is that hackers can use that extended range to access resources outside the walls of your offices.

The most common attack is called the "drive-by," where a wireless hacker is driving down the street just looking to see if there are any wireless network signals that he could either access or eavesdrop on.

One of the best ways to determine if your wireless signal extends too far is to use site survey tools which measure the range of your access point transmissions both internally and externally. You can also use these tools to assess your overall level of security and vulnerability in an effort to protect your data assets.

Site survey tools are beneficial in creating a "virtual map" of your signal coverage area. However, it is important to remember that this is only an estimated coverage map. Each vendor accounts for signal strength differently, so you must judge each result accordingly and take into account that the signal may be slightly stronger or weaker than indicated.

Special vendor settings Depending on the specific vendor of your wireless LAN equipment, it may be possible for you to set additional wireless settings that can increase or decrease your range accordingly. For example, if your signal strength is too high and you don't require an extensive coverage area, then you can adjust the power levels of your signal strength in an effort to make it less likely for a hacker to do a "drive-by" in an attempt to access or eavesdrop on your WLAN.

Directional signals In addition to adjusting the power levels to limit the range of your wireless network, you may also find it useful to use directional antenna arrays so that the entire RF signal is focused in the area where your wireless users will work. There is no need to have an omnidirectional antenna transmitting your WLAN to the corporate offices next door, since that just leaves you with a potential vulnerability waiting to be exploited at your expense.

Conclusion: Common Sense Access Controls

Maintaining control over your wireless systems in an effort to prevent unauthorized access while maintaining privacy is an attainable goal. Some of the most effective means of preventing unauthorized access are the easiest.

Since your WLAN is composed of both hardware and software solutions, you can, at the very least, evaluate your solutions by upgrading your access point configuration so that you can update your software solution and hardware firmware with the following key elements:

- Software patches
- Firmware upgrades
- Authentication routines
- Stronger encryption
- Intrusion detection systems
- Biometric access devices
- VPNs (to add another layer of encryption protection)
- Public-key infrastructure solutions

Configuration issues allow you to establish your security policy guidelines with respect to setting:

- Administrative passwords
- Encryption
- MAC screening (this only allows authorized network card access)
- Access control lists (restricting access to authorized users)

You should also remember *always* to change any default passwords for your routers and other wireless devices. Any default setting can become an extreme vulnerability that any hacker can exploit. There are even dedicated hacker Web sites that list every default password for all known wireless routers. If your router has any default access setting enabled, you can be sure that it is a simple matter for someone to figure out how to gain access just by knowing the model number and brand of your specific equipment.

Encryption settings should always be set at the highest possible values, preferably using a 128-bit level of encryption to make it that much harder for anyone to determine ways in which to eavesdrop on your WLAN.

The most common way in which a hacker enters your WLAN is when you have an "open system" enabled, where anyone in range can access your system. An easy way of stopping unrestricted wireless network access is to use medium access control (MAC) and access control list (ACL) functionality that screens out the unique ID of all machines except for those authorized to use your network.

A basic but commonly overlooked security measure is to change the default SSID of your access point. Hackers can easily log into a system whose only means of protection is a unique SSID. This information is extremely easy to acquire and can enable someone to access your system by just knowing the value of your SSID.

If the manufacturer has enabled encryption on your access point, you should immediately change its cryptographic keys because, as indicated earlier in this chapter, any default value (including encryption keys) is easy to obtain and represents a significant vulnerability in the access barrier that prevents unauthorized users from accessing your WLAN.

Most access points are preconfigured to use a specific wireless channel. This value must also be changed. In many cases, using channel 6 is often the least intrusive if you are running 2.4-GHz cordless telephones along with your 802.11b network. However, no matter what your default channel is set to, change it immediately so that you don't give any advantage to a hacker.

Finally, you should refrain from using DHCP on your wireless network because if a hacker does breach your security barrier, your DHCP server won't realize that a hacker (as opposed to an authorized user) just joined your network. The access point or wireless DHCP server will simply assign a DHCP address, making the hacker's job that much easier. With a DHCP address automatically assigned to incoming mobile devices, you are inviting intruders. Make certain that you have predefined each mobile device IP address, so that at least you can track an IP address to a given user. This gives you greater control over your WLAN and lets you keep a log of all incoming traffic so that if a wireless device is compromised, you can more effectively track the breach.

In following these guidelines, creating an effective security policy, and remaining vigilant about knowing the configuration settings of your wireless network, you can effectively prevent unauthorized access attempts into your wireless network. You can maintain an effective level of privacy and protect your mission-critical data assets from hackers.

CHAPTER 9

Open System Authentication

Using an 802.11b network on an open system opens up an entirely new set of security problems, because the authentication method used by most modern operating systems is based on using an algorithm where anyone in the vicinity of the access point can log into the network. This presents a host of security problems. Employing efficiency in connection often reduces your security.

What is Open System Authentication?

Open system authentication is the IEEE 802.11 default authentication method, a simple, two-step process, as shown in Figure 9.1.

1. The station wanting to authenticate with another station sends an authentication management frame containing the sending station's identity.
2. The receiving station then sends back a frame indicating whether or not it recognizes the identity of the authenticating station.

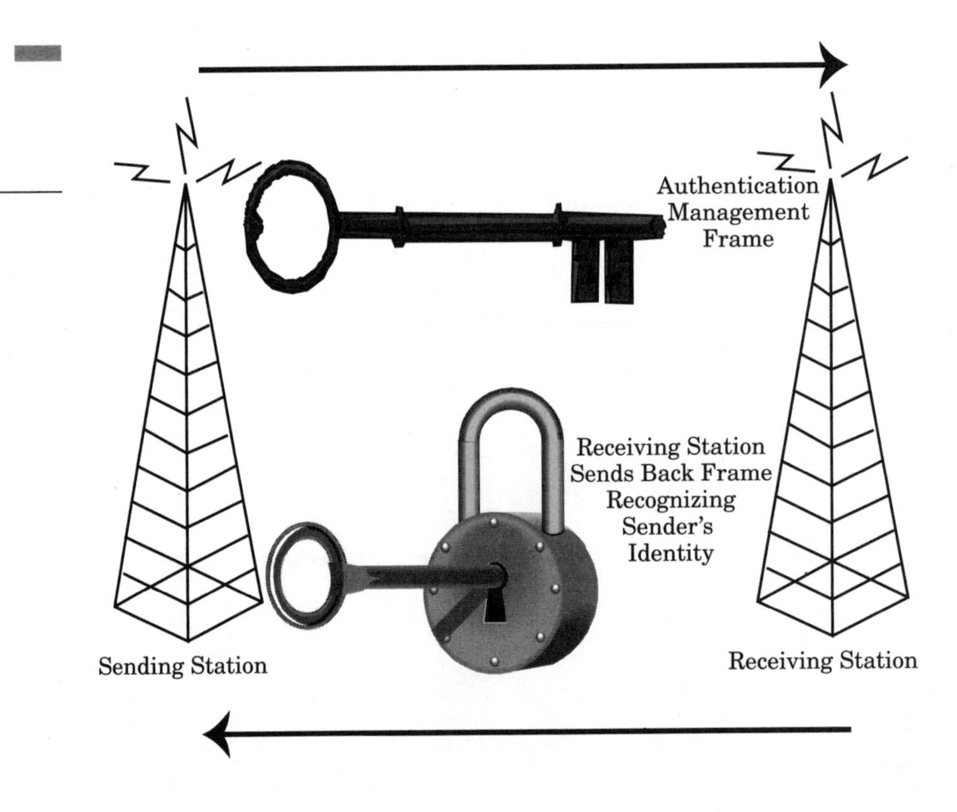

Figure 9.1 Open system authentication process.

The key concern in this area relates to the most common implementation of 802.11b in conjunction with Windows XP. The majority of users in the corporate environment will be using Windows XP for some time to come.

Windows XP has integrated support for 802.11b networks by default within the operating system. In dealing with an "open system" under Windows XP, there are several key matters to consider before deploying your WLAN.

802.11 Networks on Windows XP

When creating a Windows XP-based 802.11b wireless network, there are three primary points of consideration: user administration, key management, and security (Figure 9.2).

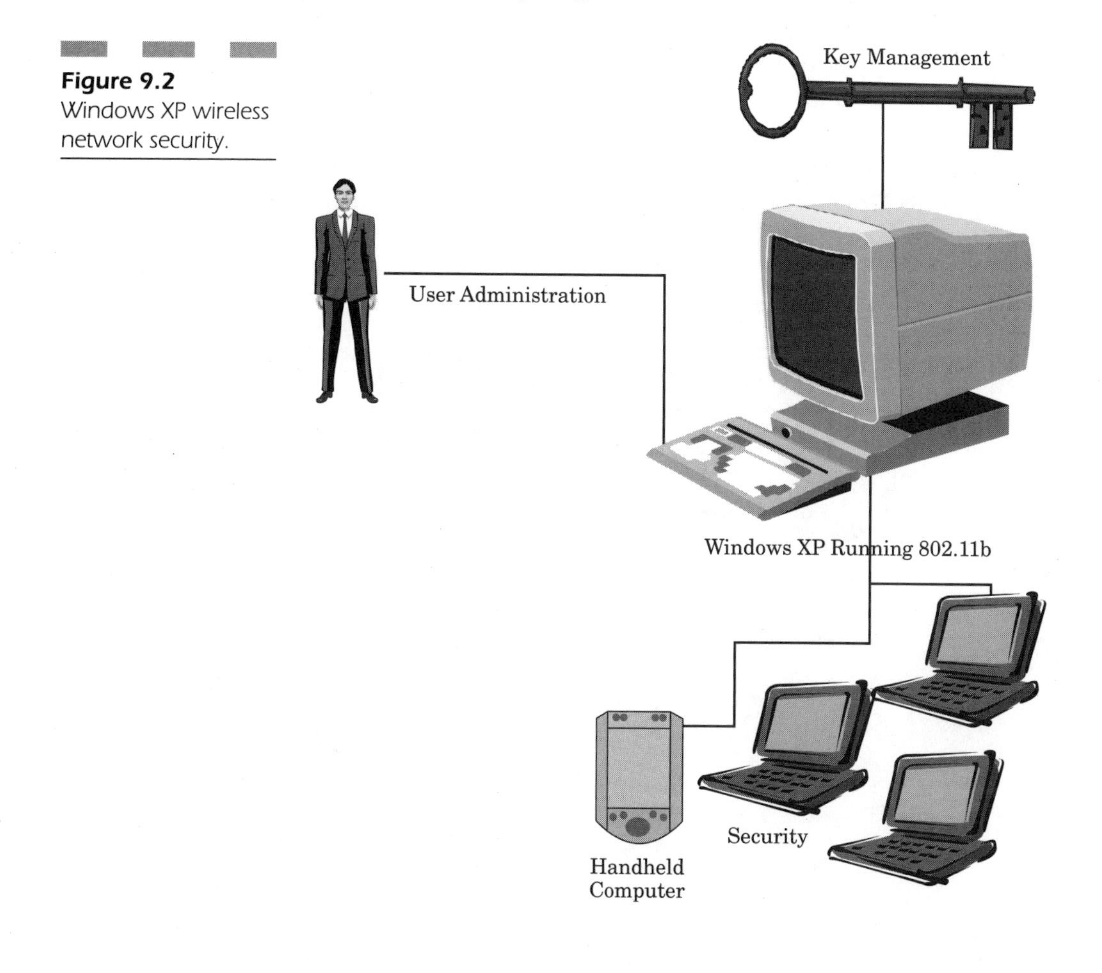

Figure 9.2 Windows XP wireless network security.

User Administration

Whenever you need to integrate user administration tools into a wireless network, there are several points to consider. Whenever you create a wireless-enabled user group, any user who is part of this group can access all resources through these wireless tools.

When administering a large network, it is important to maintain some sanity in keeping track of everyone. It is easiest to identify users through their usernames in larger wireless networks. Using the machine address of each user's individual network interface card as a means to track each user is very cumbersome. It is acceptable to restrict access based on the machine/MAC address so that you can prevent unauthorized users from accessing your network, but using that information to keep track of every user is very difficult and impractical.

When you keep track of users by their usernames, you can also check the log information on user activity to determine if there are any unusual types of hacking activities (Figure 9.3). Information you can keep track of for each username includes:

- Network usage
- Time accounting (hours of usage)
- Auditing of user activity

Figure 9.3 Logging and tracking information.

If there is a spike in network usage at any time, then it is possible that the user's identity has been stolen and that his account is being used to gain unauthorized network access. Keeping track of usage helps you determine these spikes more easily. This allows you to maintain an open system for authorized users, but a closed one for those not authorized to use your WLAN.

Time accounting is a good method of determining who should be using the WLAN and when. If you know that a user is supposed to be using the network within standard work hours, but there is an inordinate amount of usage before or after the specified working hours, then it becomes a good possibility that someone else is using that person's wireless account to gain unauthorized access to network resources. Good time accounting helps you keep track of unusual usage patterns that can constitute a network breach.

Auditing user activity helps you determine if a pattern exists that might show a breach of your network. Many intrusion detection systems audit usage logs in an attempt to determine if there is a pattern that might indicate a hacker at work. In fact, audit logs are even used by certain agencies to track down and find hackers who gain unauthorized access to your network during off hours. This process helps your administrators determine if improper activity originates from both authorized and unauthorized users.

Managing Keys in an Open System

It is very difficult to manage encryption keys that never change. Whenever you leave the same key on your wireless station or access point for any extended period of time, it becomes highly vulnerable to being hacked. It is important to use a unique method of managing your keys or at least storing them in databases that are not necessarily connected to your network.

Authentication Concerns

In the 802.11b environment, it is important to note that there is no per-packet authentication mechanism. This means that you cannot analyze the packet level to determine if any given packet of data transmitted across your WLAN is being corrupted by someone trying to destroy the validity of your data or cause interference on your network.

You are still vulnerable to disassociation attacks with 802.11 associate/disassociate messages that are unencrypted and unauthenticated. This could allow forged disassociation messages to be used against clients. Your best defense under these circumstances is to add a keyed message integrity check (MIC).

In an open system authentication there are no levels to protect your network. This means that someone could easily log into your network without a user identification or authentication. Furthermore, there is no central point of authentication, authorization, or support for accounting types.

Even though you might believe that having an encryption cipher in RC4 will protect you, it is important to know that it will not offer you protection against plain text types of hacker attacks.

We have discussed WEP keys, but many systems are vulnerable to having their keys reverse engineered just because user passwords are known. This can effectively negate any type of WEP protection you have on your network and leave you quite vulnerable to an attack in which a hacker eavesdrops on your network connection and determines ways to decipher the mission-critical data on your WLAN as well as on your wired LAN (Figure 9.4).

Another problem is that there is no support for any method of extended authentication that includes:

- Public/private-key certificates
- Smart cards
- One-time passwords
- Biometric authentication devices
- Token cards

There is no method of managing dynamic unicast session key (as opposed to a multicast global authentication key) for each wireless workstation. Such issues involving key management and the rekey of global keys are a known weakness in many WLAN implementations.

802.11b Security Algorithms

The algorithm in 802.11b offers few security options to keep your data protected against a persistent hacker. 802.11b is an open system by default, meaning that authentication and encryption servers based on the WEP algorithm are not activated by default. Even when they are

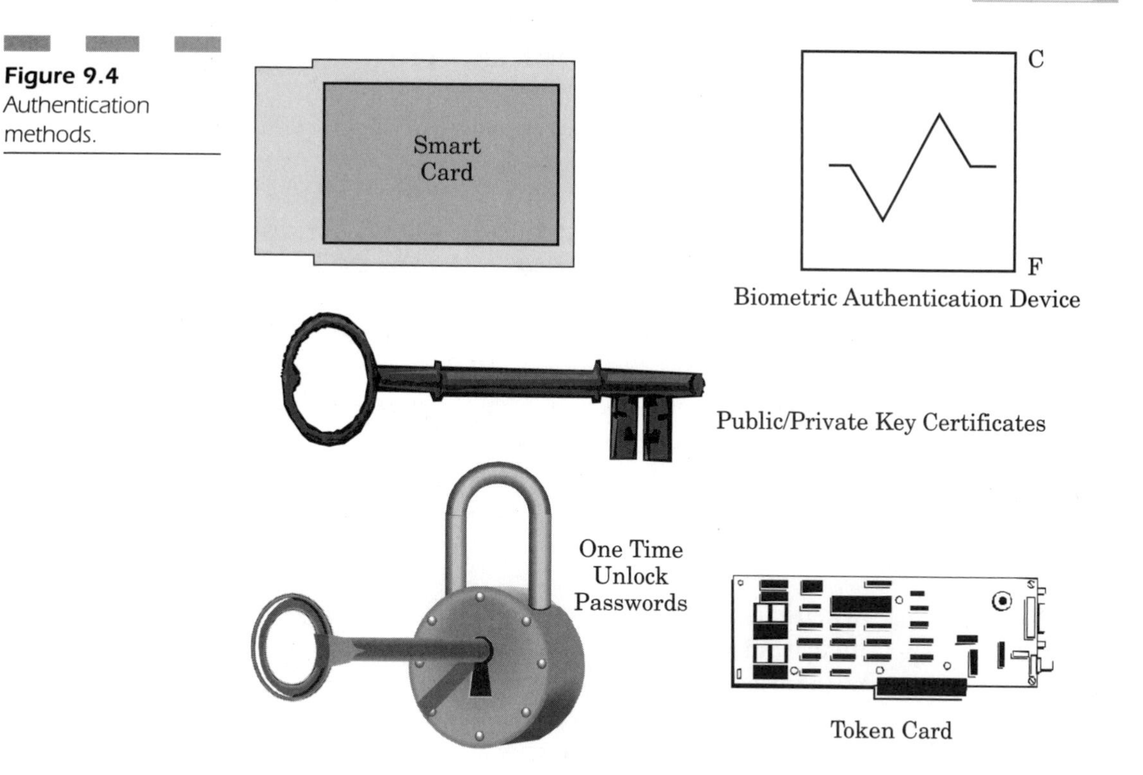

Figure 9.4 Authentication methods.

activiated, given enough time, a hacker can break this algorithm and force access into your WLAN.

Authentication Support

The two major types of authentication support within the 802.11 standard are open system and shared key.

These types of authentication methods involve authentication controlled by a parameter known as "authenticationtype." The actual type of authentication that the wireless workstation accepts is controlled by the security management information base (MIB).

An "open system" is the default "null type" of authentication algorithm that utilizes a two-step method composed of 1) identify assertion and 2) request for authentication.

The result of this process is the authentication result. This means that the user is correctly authenticated as an authorized network user via a protocol that protects the transmission of data from the wireless workstation to the access point.

Shared-key Authentication

Shared-key authentication supports wireless workstation authentication methods in one of the following ways:

- Member of a group that knows a shared secret key
- Member of a group that does not know the key

The 802.11 standard automatically makes the determination that the shared key is provided to the wireless workstation via a secure channel that is completely separate from the wireless channel used by 802.11 for data transmission.

Secret Keys

When dealing with secret keys, you are really working with the WEP algorithm. In most cases, you have 40-bit secret keys used for both authentication and encryption. Most 802.11 implementations permit higher-level encryption when using 104-bit secret keys. Although 802.11 does not force you to use the same WEP keys with all wireless workstations, it does permit each wireless user to have sets of shared keys:

- Unicast session keys
- Multicast or global keys

Most 802.11 setups support shared multicast/global keys; however they will shortly support unicast session keys for each wireless workstation.

You will find that you have encryption services from WEP that are used to protect authorized WLAN users from hackers who are trying to eavesdrop on network traffic. However, WEP allows your WLAN to emulate the same types of physical security attributes present in your wired LAN, as long as you have taken the safeguards and precautions outlined in this book to prevent your personal information from falling into the hands of someone trying to break into your network.

The WEP Algorithm

The key used for both encryption and decryption is based on a symmetric algorithm used by WEP. The secret key is combined with an initialization vector (IV) and produces a component that is used as an input to a pseudo-random number generator (PRGN); this results in a mathematical key sequence linked with the message text and an integrity check value (ICV).

Essentially, you are dealing with three primary components that produce the 802.11 data frames:

- Initialization vector (IV)
- Actual message text
- Integrity check value (ICV)

Static Vulnerabilities

When you keep the same secret key static for too long, the IV will be modified every so often with each MAC protocol data unit (MPDU). The frequency at which the IV values are modified depends on the level of privacy needed by the WEP algorithm. If the IV is changed after each MPDU, you are in the best position to keep WEP intact on your WLAN.

One of the greatest problems in 802.11 security is that there are a number of difficulties with the key management protocol in WEP. When dealing with a wireless workstation, the problem is more pervasive. Furthermore, there is no good implementation of either authentication or encryption services when working in in "ad hoc mode."

The security options for access control do not work as well as they should for larger-scale network infrastructures because there is no interaccess point protocol (IAPP), which makes key management much more problematic when wireless stations roam between access points along your network.

NIC Security

Network interface cards (NICs) work much more efficiently if you can eliminate as much complexity as possible when dealing with both infrastructure and ad hoc network modes. What you need to do is set the network adapter configuration through an automated method, because the biggest problem is when the wireless user starts setting these parameters himself.

Configuration problems that need to be fixed usually deal with client configuration, most especially when working with multiple operating systems. Whenever the client moves between one operating system and another, you can be sure that you will have to reset an entire set of configuration options for the user and even some of the network resources that he accesses.

Most 802.11 NICs support default methods of authentication. The default authentication algorithm first tries to use shared-key authentication if the network adapter has been preset to use a WEP shared key. However, if the level of authentication stops working because the NIC is not set up with a WEP shared key, then the NIC will always go back to its lowest common denominator, open system authentication. This opens the floodgates for unauthorized users to roam right onto your network virtually undetected.

Wireless NIC Power Settings

Wireless NIC cards are powered in two ways:

- Desktops plugged into the wall
- Laptops running on batteries

Standard settings Machines often have a client name that distinguishes them on the WLAN, and in most cases this name is set by default as the machine name. This situation is very insecure and leads a hacker right to your wireless workstation. That is why it is very important to use a nonstandard means of identification for each wireless workstation so that a hacker cannot know which machine is which; the administrator is the only one who truly knows. This adds a level of privacy that also acts as a small security mechanism; you should not give away too much information about your network to anyone who doesn't have a *need to know*!

Media connection events Some of the items that hackers look for in any open wireless network involve the way in which the wireless NIC supports *media sense*. An event occurs whenever the media connects with a new access point. In fact, a disconnect event is not even necessary unless the NIC has totally lost its connectivity to the wireless router. Note that any connection event indicates to the transport layer that it should be aware that there might be a transition from one subnet to another.

Open System to WEP Authentication

The major problem inherent in an open system is that anyone can potentially eavesdrop on everything you transmit from the wireless workstation to the wireless router/server. The best defense against open system problems is to migrate to WEP authentication, thereby giving yourself protection in many ways equivalent to using a wire-based LAN system.

When you have several access points configured with the same WEP key, it is important to note that your access point can use another form of optimization. The wireless NIC card first attempts to execute 802.11 authentication using the WEP key acquired by the older access point and uses that value as the shared key. If that method works, then the access point usually adds that wireless workstation to the authenticated access list and allows that user to access all the resources of the internal corporate network.

However, if that authentication does not work, then the wireless NIC card will use the open system authentication method to communicate to the access point, thereby authenticating that wireless workstation to the WLAN.

The access point's job is to determine if that wireless workstation used an open system authentication or execute a shared-key authentication method when logging into the WLAN. If that wireless workstation acquired access to the new access point using shared-key authentication, then the 802.11 authentication will be started by the new access point to update its logs concerning that wireless user.

When you allow a wireless workstation to connect using shared-key authentication, the new access point makes certain that the wireless workstation does not experience any problems with its network connectivity. Should the wireless workstation not be able to authenticate itself correctly to the new access point, the wireless workstation network connectivity through the access point controller port is stopped in an attempt to maintain network security.

Port-based Network Access Control

Port-based access control enables authenticated network access for local area Ethernetworks. It uses the physical components of a switched LAN network so it can offer a method of authenticating devices connected to a specific LAN port. This method effectively prevents access to that specific port when there is no successful authentication.

A port access entity (LAN port) can take on specific roles with respect to access controls, as authenticator or as supplicant.

The "authenticator" is the port that makes certain all entities are authenticated before permitting access to services that can be accessed on a given port. The authentication server (which can either be a separate unit or have its functions within the authenticator) executes the authentication method to inspect the "supplicant's" credentials for the authenticator. It then replies to the authenticator to determine if the supplicant is authorized to access the authenticator's services (Figure 9.5).

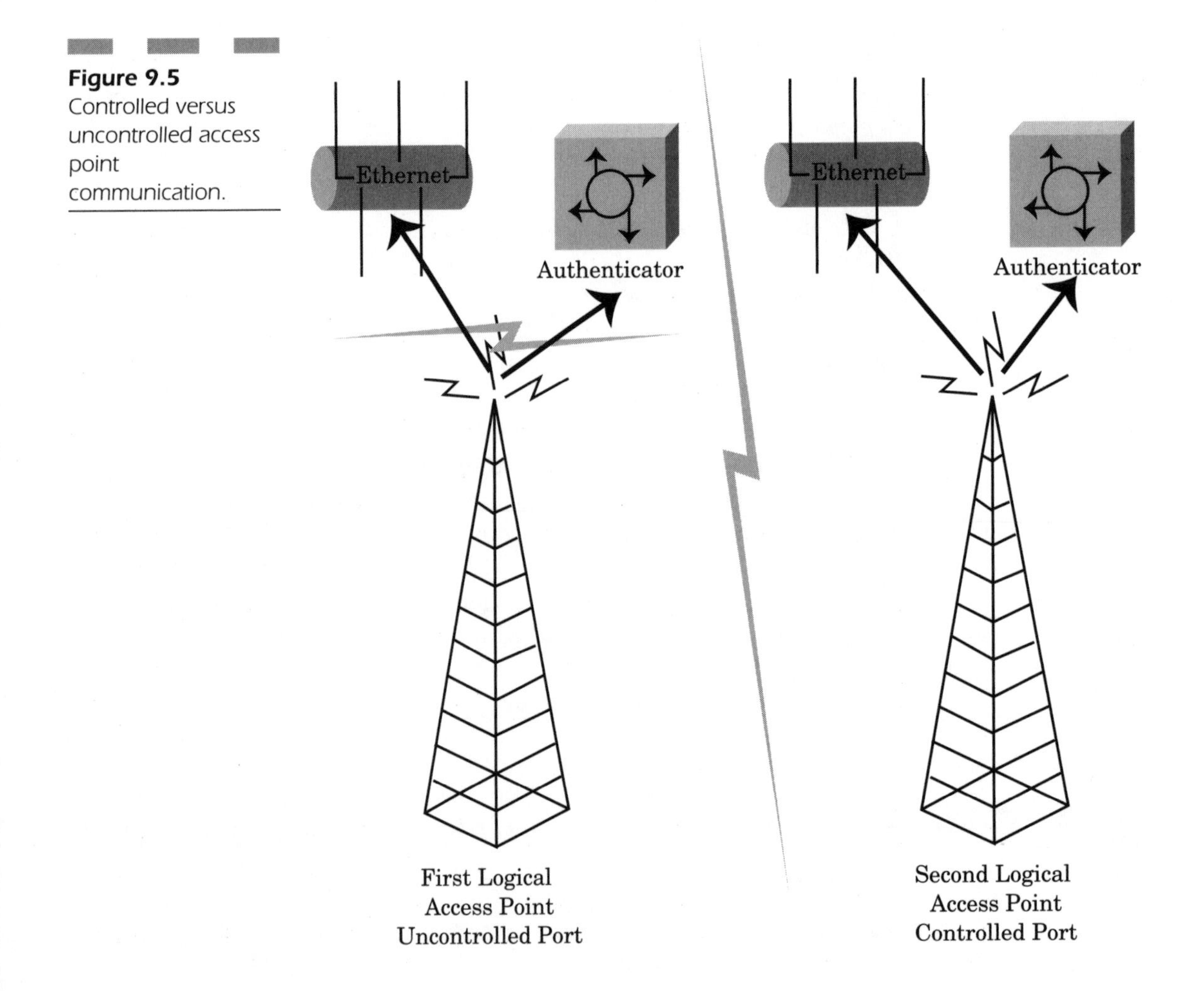

Figure 9.5 Controlled versus uncontrolled access point communication.

Port-based access comes into play with the authenticator with respect to two logical access points to the LAN through one single LAN port:

1. The logical access point is an uncontrolled port that permits an uncontrolled exchange between the authenticator and the other LAN systems. This occurs irrespective of the system's authorization.
2. A second logical access point is a controlled port that allows communication between the LAN system and the authenticator services. This happens only when you are dealing with an authorized system.

Securely Identifying Wireless Traffic

The 802.11 standard must permit a wireless access point to identify traffic securely for specific types of clients by sending an authentication key to the client as well as to the wireless access point; this is the default authentication procedure. Only authenticated clients actually know the authentication key, and that the same key will encrypt all packets transmitted by the client. If there is no valid authentication key, then the "authenticator" will restrict wireless traffic passing through it. On the other side of the coin, when the wireless workstation or "supplicant" is in range of the access point, the access point sends a challenge back to the wireless workstation. When the wireless workstation receives the challenge from the access point, it transmits its identity back to the access point, which then sends the identity of the wireless workstation to the authentication server to begin the authentication process.

At this point, the authentication server then asks for the credentials of the wireless workstation. It then determines the types of credentials it specifically needs to confirm the wireless user's identify. Note that all the requests sent between the wireless workstation and the authentication server go through the uncontrolled access point port so that the wireless workstation is not able to contact the authentication server directly. In addition, the access point does not permit responses through the controlled port because the wireless workstation does not have the required authentication key.

The wireless workstation then sends its credential to the authentication server and, upon validation, the authentication server sends an authentication key to the access point. That key is encrypted, so that only the access point has the ability to send. The access point can use the authentication key it got from the authentication server to transmit securely to each wireless workstation with both a unicast session key and a multicast/global authentication key.

Extensible Authentication Protocol

The extensible authentication protocol (EAP) is required to encrypt the global authentication key. EAP offers a method necessary for wireless workstations to be able to create an encryption key for the authentication service.

Mutual authentication is provided by transport level security (TLS) to protect the integrity of encrypted transmissions and the exchange of keys from point to point. Because a combination of EAP and TLS is used, TLS mechanics facilitate EAP.

Once authentication has occurred, 802.11 can be set to request that the wireless workstation authenticate itself again at a predefined time interval. This means that the wireless access point is set to restrict network traffic when it is sent to a wired network or other wireless workstation without valid authentication keys.

Both the wireless access point and wireless workstation need to support a multicast/global authentication key so that the wireless access point can utilize a server that receives 802.11 network traffic either with or without a specific authentication key.

When the access point has a new wireless workstation connecting to it, the access point receives an EAP-Start from the wireless workstation. Then, the access point sends an EAP-Request to the wireless workstation, to establish its identity. The access point then sends an EAP-Start connected with the new access point on your WLAN.

The wireless workstation can then send an EAP-Response using as an identifier the same specific machine name as the response request if there is no user logged on at the time. The wireless workstation can send an EAP-Response using as an identifier the same username as that request if there is a user logged on at that time.

At that point, the EAP-Response for identity is sent by the access point to the authentication server, which then transmits an EAP-Request via a TLS or MD5 challenge to the EAP-Response for an identity message from the wireless workstation.

Note that TLS is necessary for wireless traffic, since the authentication server is not able to permit sending multicast/global keys. The wireless workstation must therefore deal securely with unicast session authentication keys so that the wireless access point sends the EAP-Request from the authentication server to the wireless workstation.

The wireless workstation then sends an EAP-Response containing its credentials to the authentication server through the wireless access point, which then sends the wireless workstation's credentials to the authentication server. The authentication server validates the wireless workstation's credentials and creates a "Success" message for the wireless workstation.

The authentication server responds to the wireless access point with the wireless workstation message and the encryption key from the EAP-TLS session key.

At that point, the wireless access point creates a multicast-global authentication key either by producing a random number or by choosing it from a predefined setting. Once the authentication server receives that message, the wireless access point sends a "Success" message to the wireless workstation. The wireless access point then sends an EAP-Key message to the wireless workstation that has the multicast/global authentication key encrypted through the per-session encryption key.

Should the wireless access point and wireless workstation support this type of unicast session key, then the access point uses that encryption key (sent by the authentication server) as the unicast session key.

Once the wireless access point alters the multicast/global authentication key, it can produce EAP-Key messages that have the new multicast/global authentication key encrypted with specific wireless workstation unicast session keys. The wireless access point then adds the specific wireless workstation unicast session key to the list of unicast session keys it has logged.

Once the wireless workstation has received the EAP-Key message, it uses the unicast session encryption key to decrypt the multicast/global authentication key. Once the wireless access point and wireless workstation receive these unicast session keys in combination with a multicast/global authentication key, the encryption key (from the EAP-TLS session key) is sent to the wireless workstation as the unicast session key to use.

Finally, when the wireless NIC receives these authentication keys, it must program the wireless workstation's NIC to accept them. When the authentication keys have been successfully programmed, the wireless workstation uses DHCP to restart its process of communication and assign an IP address for itself.

Conclusion: Open System versus Closed System Authentication

This chapter deals with the problems associated with having an open system of authentication in your WLAN. While it may be easier to deploy and simpler for users to connect, it presents a terrible risk in your security that could leave your system open to an attack by a hacker and make it easy for someone to compromise not only your wireless network, but your entire intranet as well.

To best defend yourself, you can utilize all the types of encryption standards inherent in an 802.11 protocol like WEP. The idea of this system is to close your open system sufficiently so that only authorized users can access your network resources. Encrypting your data also protects your network traffic from prying eyes trying to determine how to intercept your mission-critical data. The concept is to make your WLAN have a level of security analogous to that of a wired LAN. In theory, this is a useful idea; in practice it is not usually accomplished because of the great number of ways in which your wireless network is vulnerable to a hacker attack.

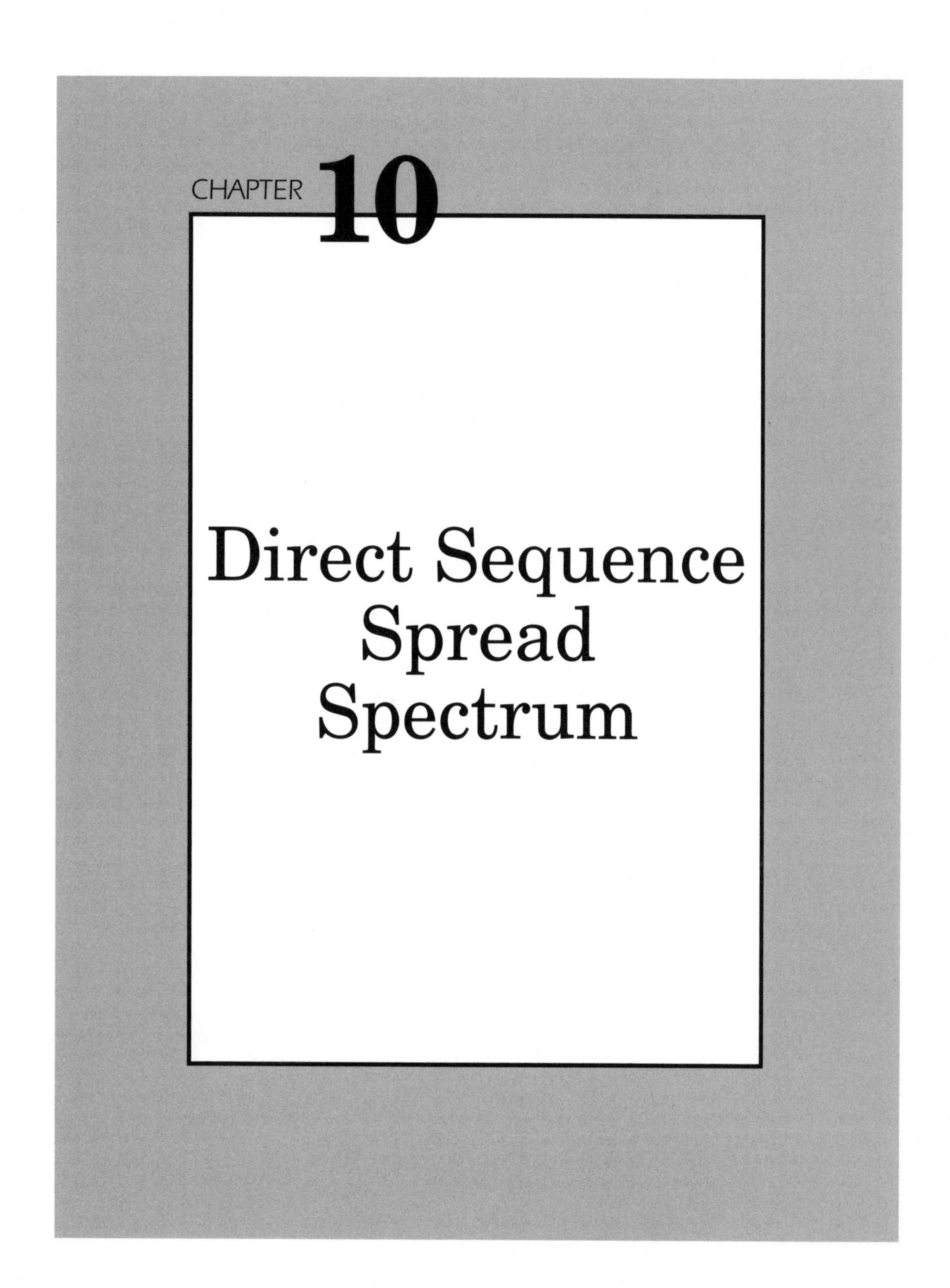

CHAPTER 10

Direct Sequence Spread Spectrum

This chapter explains how 802.11b DSSS is static in frequency and also uses a single DS "spreading code" for all time and all users. Anyone desiring to do so can generate valid 802.11b control packets which must be accepted by all 802.11-compliant equipment; alternatively, anyone can listen to all 802.11b control frames transmitted. (The complexity of wireless data-link protocols makes the comprehensive enumeration of specific denial-of-service attacks impossible.)

802.11 DSSS

In order to achieve high-speed wireless data networking, 802.11 was created to foster interoperability between various brands of WLANs. The goal was to create a "universal technology" that was platform independent and both provided higher performance and interoperability using products from different vendors. This permits wireless users to use any hardware solution mix necessary to satisfy application requirements.

Standardization

Making 802.11 an industry standard provides for a decreased component cost for users so that you can implement a WLAN cost effectively. The 802.11 standard permits you to choose equipment that offers direct sequence spread spectrum (DSSS) or frequency hopping spread spectrum (FHSS), both based on radio frequency (RF) transmissions.

802.11 started out by having DSSS support different physical layers (PHY) at a 2-Mbps peak data rate that can fall back to 1 Mbps in very noisy areas. FHSS PHY functions at 1 Mbps and permits 2 Mbps in open environments without any interference.

The evolution of 802.11 allowed the implementation of DSSS at higher data transmission rates of 11 Mbps, making the transition from the 2-Mbps 802.11 DSSS system to a system at 11 Mbps simple because the modulation methods are analogous. In fact, 2-Mbps DSSS systems will operate alongside 11-Mbps 802.11 systems to provide a seamless transition between lower and higher rates of data transmission. This is much the same as moving from 10-Mbps to 100 Mbps wired Ethernet in an effort to allow greater performance enhancements without having to revamp the entire protocol to make things work together.

MAC Layers

The media access control (MAC) layer is powerful, with enough features to support sequence control as well as Retry fields that support "MAC layer acknowledge," which reduces interference and increases the usage of available bandwidth on a given wireless channel.

In order to ensure reliable communications when other stations are present, you need the following MAC fields (Figure 10.1):

- Type
- Subtype
- Duration
- WEP (wired equivalent privacy)
- Sequence control
- Frag

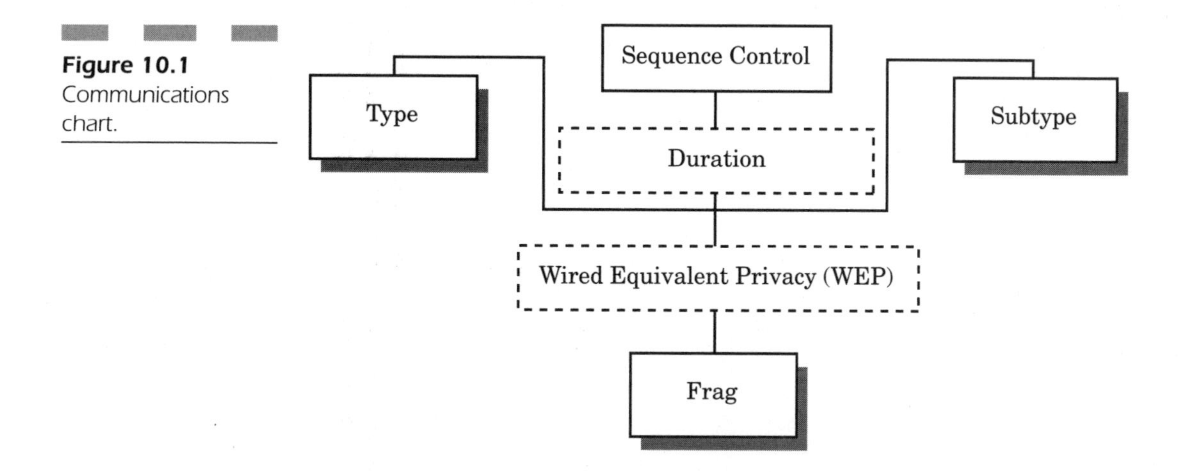

Figure 10.1 Communications chart.

WEP fields permit data security that is analogous (in some respects) to the physical security characteristics of a wired Ethernet. Both sequence controls and Frag fields deal with "fragmentation," which permits a WLAN to function in tandem with devices that cause signal fading or interference patterns.

MAC works very easily with normal wired Ethernetworks in combination with either an access point or a wireless router. The idea is to make certain that wired and wireless nodes on your LAN can function seamlessly with each other.

CSMA

WLANs use a standard referred to as *carrier sense multiple access* with collision avoidance (CSMA/CA) as a MAC method. However, normal Ethernetworks use a *carrier sense multiple access with collision detection* (CSMA/CD) method.

Roaming

Regardless of what equipment you use, 802.11 permits a wireless client to roam across multiple access points (Figure 10.2). These access points can function on either the same or different channels. After a certain interval has elapsed, an access point may transmit a beacon signal (with time stamp) to execute the following tasks:

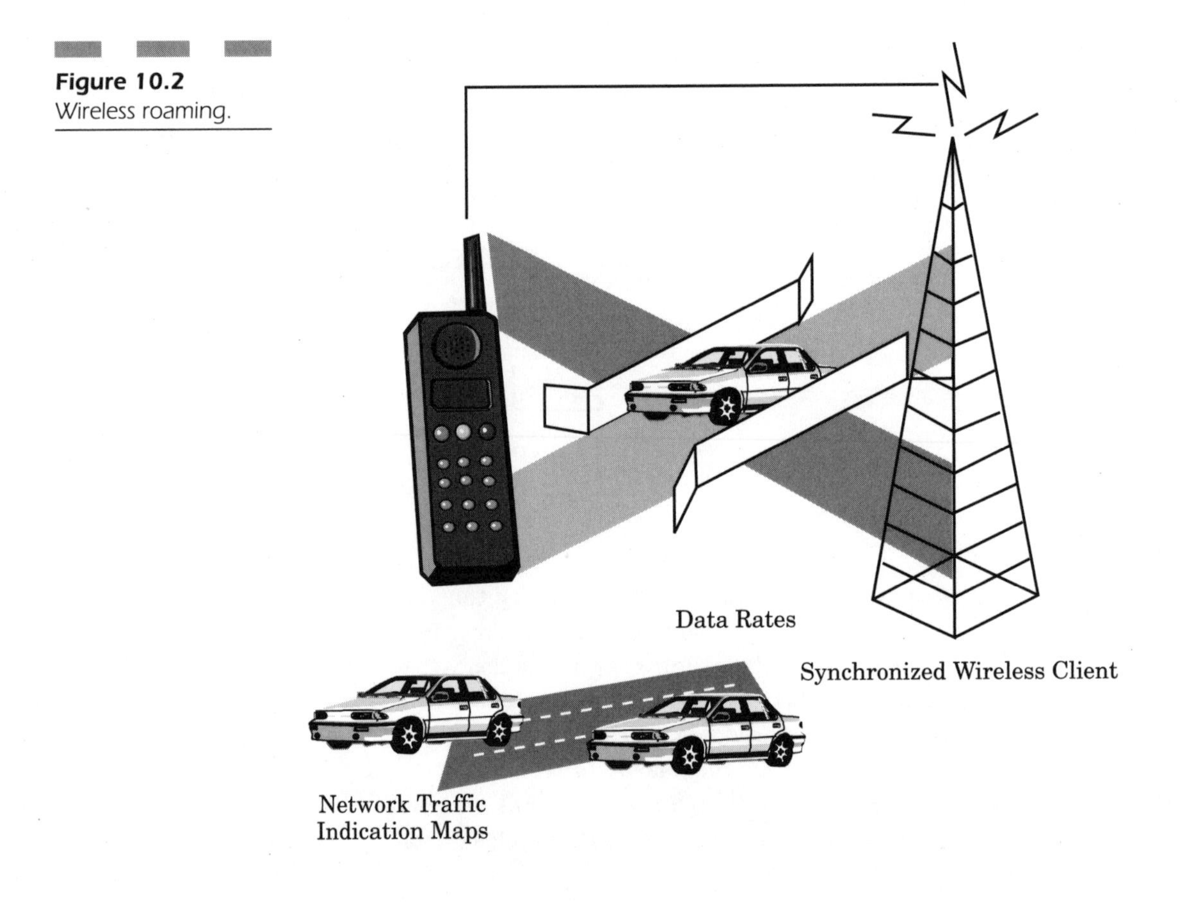

Figure 10.2 Wireless roaming.

- Synchronize wireless client
- Indicate supported data rates
- Indicate other parameters
- Provide a traffic indication map

When a client roams, it uses the transmitted beacon to determine the strength of its existing connection to the access point. Should the connection appear to be weak, then the roaming station can try to link up with another access point to sustain its connection to the network.

Power Requirements

One of the advantages of using DSSS is that it is important to conserve as much power as possible with wireless PDAs and other battery-operated remote connection devices on your wireless network. Unless you have sufficient battery life in your device, your device may shut down after only an hour or so of use.

The 802.11 protocol has enhanced MAC features to increase battery life through specific power management methods. Unfortunately, power management schemes cause difficulties with WLANs because standard types of power management methods derive their savings from placing the wireless device into a "sleep mode" that basically turns the unit off. When there is no network activity for a specific amount of time, the unit is not able to receive important data transmissions.

In order to support wireless clients that are put into sleep mode, 802.11 makes it possible for access points to include buffers designed to queue messages. This means that sleeping clients must be awakened every so often to receive important messages. However, access points are allowed to trash unread messages after a certain amount of time has elapsed so that obsolete messages do not remain on the server.

Increasing Data Transmission

As 802.11 evolved, rates of data transmissions increased to 11 Mbps early in the process of ratifying the specification. The 11-Mbps PHY layer uses *complementary code keying* (CCK). This standard is based on DSSS and offers speeds up to 11 Mbps (Figure 10.3). However, as distance increases between the wireless user and the access point (or if there is interference) the rates fall back to various ranges including:

- 1 Mbps (best)
- 5.5 Mbps (very good)
- 2 Mbps (good)
- 1 Mbps (fair)
- 0 Mbps (out of range)

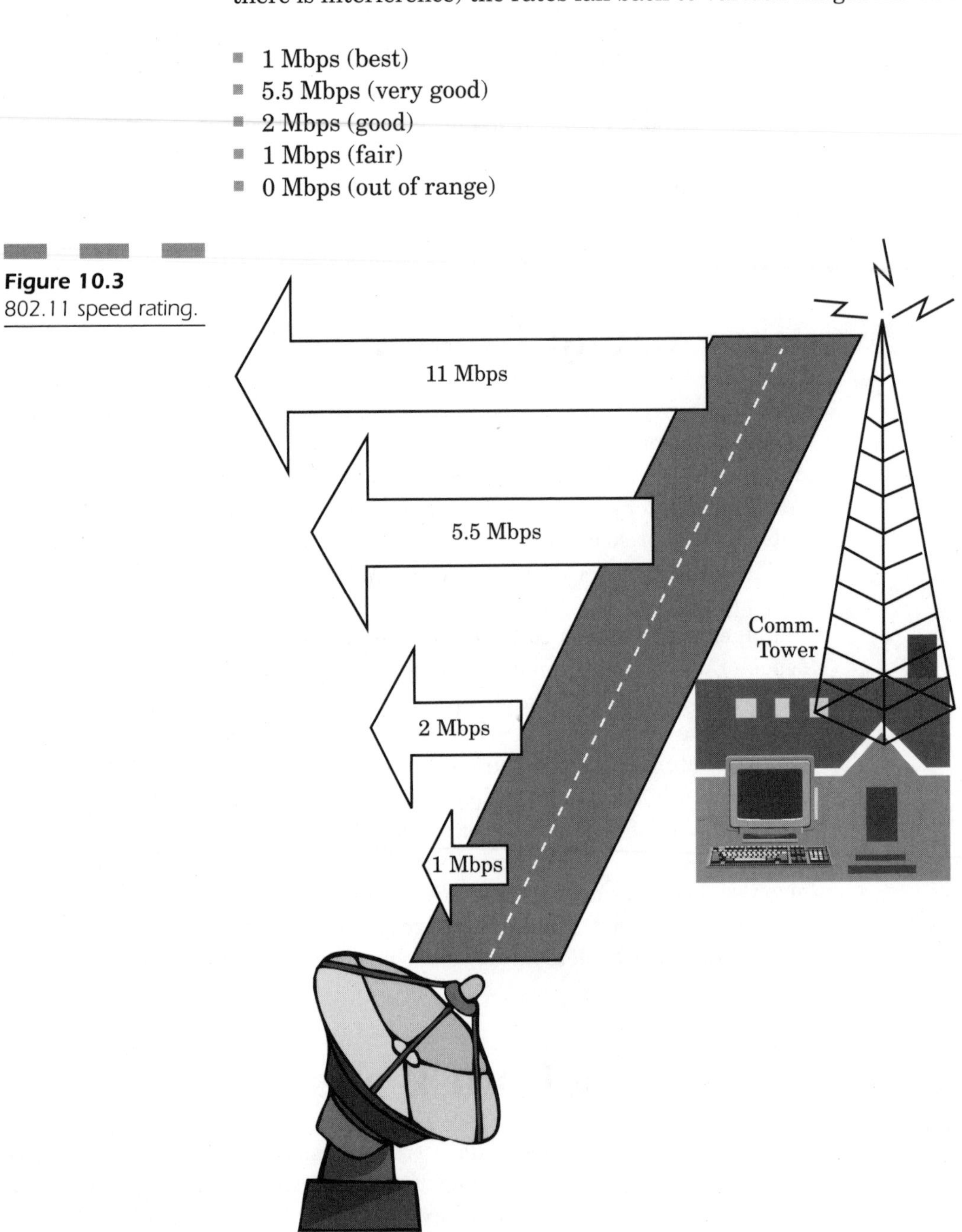

Figure 10.3
802.11 speed rating.

Because standardized wireless devices all adhere to the 802.11 standard, all data rate ranges can be supported, even slower, legacy DSSS systems. In contrast, when dealing with wired Ethernets, higher speeds are necessary to keep up the pace with broadband applications that require increased bandwidth for such items as shown in Figure 10.4:

- Streaming video and audio
- Internet telephony (VoIP)
- Multimedia applications
- Installing network-based applications

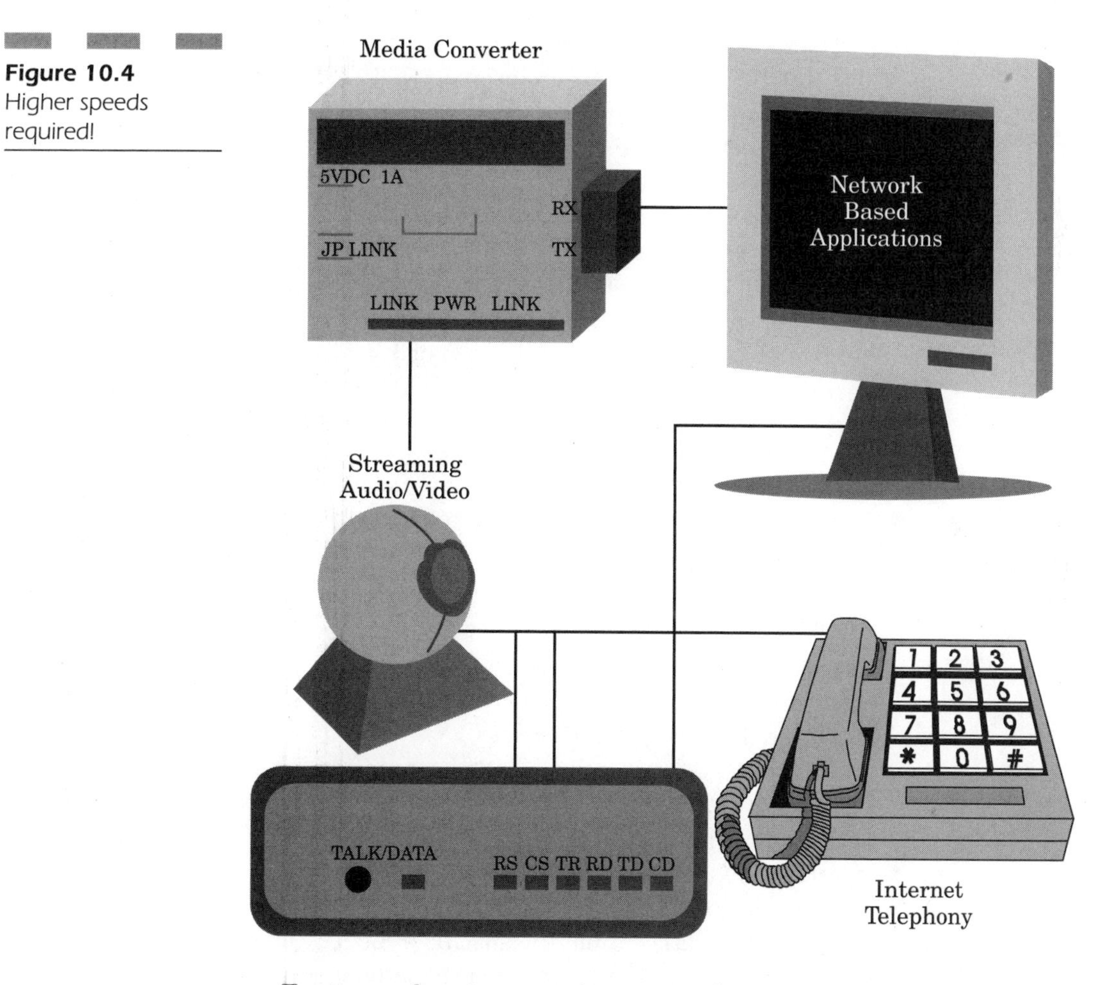

Figure 10.4 Higher speeds required!

Faster peak rates permit more nodes to connect efficiently to your WLAN through one channel. In addition, vendors are proceeding with new 802.11a applications that have speed increasing from 11 Mbps to as fast as 54-Mbps in the 5-GHz band.

FHSS Security

One of the most pressing problems in WLANs is the question of whether or not frequency hopping can increase the security of your wireless network. You will notice there are a number of people who tout the security of HomeRF using FHSS (as opposed to DSSS) in 802.11b networks.

HomeRF proponents insist that frequency hopping makes it far more difficult to eavesdrop on or intercept network traffic. In addition, it is difficult to decipher this information, which is transmitted all over the spectrum. 802.11 using DSSS is said to be more susceptible to these types of security concerns (eavesdropping and interception) because it uses the same channel to transmit both data and security information—making it easier for someone to circumvent the inherent security measures of the 802.11 protocol.

However, there is no “real” benefit to HomeRF over 802.11b with respect to security issues. All types of WLANs support distinct types of security protocols; both FHSS and DSSS systems employ methods of data encryption to stop any types of unauthorized eavesdropping of network traffic. Furthermore, the user authentication procedures of 802.11b stop unauthorized hackers from acquiring access to mission-critical data.

In many cases, it seems that FHSS offers a superior level of security because of the design elements of this transmission technology. While there are some elements that could make FHSS more secure than DHSS, the principal element that gives it greater security includes “hop sequences” that are specified by somewhat unpredictable methods of spectrum usage. Hop sequences are generated by HomeRF radios are designated in five seconds or less.

HomeRF systems utilize FHSS modulation in an effort to satisfy the regulation set by the FCC with respect to radio operation in the 2.4-GHz ISM band. The idea is to make these networks comply with regulatory specifications rather than to provide security.

HomeRF networks do not have any security mechanisms to prevent hackers from determining the specific frequency hop set their devices use; what is supposed to be a more secure method is essentially less secure.

Even the algorithm used for hopping is not necessarily one of the actual elements controlling security; the HomeRF control point (access point) sends the hop-set identification information unencrypted across the network from every beacon. This action takes place each time the network hops channels (as many as 50 times per second). Should the hop-set identification information be sent across the network unencrypted from the control point beacon, the hop set could still be deciphered.

Hop Sequences

FHSS radio transmissions, by definition, change their operating frequency according to a semirandom pattern. Due to the random method of the hop sequence, it is somewhat protected against hackers trying to eavesdrop on network traffic. However, with HomeRF, the hop sequence is deciphered in less that five seconds because the hop is somewhat slow, at only 50 hops per second. By comparison, Bluetooth is considered slow and its speed is far in excess of FHSS at 1600 hops per second.

Additionally, there are only a small number of different hop patterns designated for HomeRF radios, in which each hop is composed of 75 distinct frequencies, with each hop repeating itself every 1.5 seconds. The specific patterns for the HomeRF specifications can be easily read from the SWAP specification by anyone interested in getting a copy of that spec.

A beacon is sent each time the network hops to a new channel in the HomeRF protocol. In fact, a hacker can eavesdrop on the beacon for only a few seconds before the hop set of a HomeRF radio can be deciphered. Furthermore, if the beacon were encrypted, you could still detect the radio transmissions and simply measure the amount of time of reception. This information alone would allow the hop set to be deciphered.

You can decipher the hop set for a standard Home RF system using 75 channels, but you can determine even more easily the hop set for *wide band frequency hopping* (WBFH) systems because they use only 15 channels. This means that FHSS systems do not have any true advantage over DSSS when it comes to built-in security features and functionality.

FHSS versus DSSS

There are reasons why some WLANs use FHSS rather than DSSS and vice versa. When you need to transmit, it is important to spread out the energy of the signal to reduce interference to other users in the radio spectrum you are using. As a result, FHSS was used by many vendors in the 2.45-GHz ISM band who were using power levels greater than 1 mW since it provided a reasonable level of "inherent" security.

Any systems using either FHSS or DSSS are permitted to transmit power up to 1000 mW, so they have sufficient power for WLAN connectivity. The algorithm that specifies the hop sequence for HomeRF is published and available through the SWAP specification; and the hop sequences are used more for regulatory compliance than for increasing security.

FHSS systems use "frequency agility" to satisfy regulatory requirements. The HomeRF beacon sends the hop-set identification information unencrypted in each CP beacon. If the hop-set identification information were not transmitted in clear text, it could still be easily deciphered just by eavesdropping on the traffic on each hopping channel. HomeRF has an FHSS system of frequency agility that does not have any security advantages over a DSSS system, regardless of the hype any analyst might tell you.

Frequency Allocation

The 802.11 offers two types of PHY layers, each with distinct RF usage through either FHSS or DSSS. Both FHSS and DSSS options were created to adhere to regulatory rules set by the FCC to operate in the 2.4-GHz ISM. The unlicensed ISM band is allocated slightly differently worldwide. Table 10.1 shows that the actual break in the spectrum of usage varies by country.

TABLE 10.1 Spectrum of Usage Varies by Country

Country	Frequency
United States	2.4000–2.4835 GHz
Europe	2.4000–2.4835 GHz
Japan	2.471–2.497 GHz
France	2.4465–2.4835 GHz
Spain	2.445–2.475 GHz

Both FHSS and DSSS support 1 and 2 Mbps, but 11-Mbps radios utilize DSSS. In fact, DSSS setups utilize the same technology as global positioning system (GPS) and satellite cell phone equipment.

The specifications of this technology require that each information bit is linked through an XOR function that has a long numerical value or a pseudorandom numerical value (PN) that produces a high-speed digital frequency modulated spectrum on a carrier frequency using *differential phase shift keying* (DPSK).

When a DSSS signal is received, it is matched to a filter correlator that removes the PN sequence and regains the original data stream. The data rates of 11 Mbps and 5 Mbps are achieved only when DSSS receivers use different banks of correlators and PN codes in order to recover the transmission stream of network data.

The high-speed rate modulation mechanism is designated as a complementary mechanism.

The PN sequences actually spread the data stream transmission bandwidth of the signal, which defines its mechanism as spread spectrum. The objective is to reduce power, and the total power used remains the same. When the signal is received, it is correlated with the same PN sequences so it can reject any narrowband interference and reassemble the binary data in its original form.

The exact speed is not as important as the fact that the transmission uses about 20 MHz for DSSS systems. This means that the ISM band can support as many as three non-overlapping channels.

The fundamental methods that 802.11 uses involve the *distributed coordination function* (DCF). It then uses *carrier sense multiple access with collision avoidance* (CSMA/CA). This means that the wireless workstation must listen for other users on the network. The station then transmits once the channel is idle; however if it is busy, the wireless workstation pauses until the transmission stops and executes a random backoff until it can transmit safely on the radio spectrum.

The space of time between the packet transmission and the start of the "ACK" frame is one *short interspace* (SIFS). The ACK frames have a higher priority than other network traffic, requiring fast acknowledgment since ACKs need to be supported by the MAC sublayer in the 802.11 standard.

Some transmissions wait for at least one DCF interframe space (DIFS) prior to sending any data across the network. Should the transmitter perceive that the network is very busy, it can then decide a specific random backoff period by determining a value for the internal timer for a specific number of slot times. When DIFS expires, the timer starts to decrease. When the timer approaches zero, the station can then begin transmitting. Should the channel be in use by another wireless workstation prior to the timer's approaching zero, the timer setting is kept the same at the decreased value for each future transmission across the network. The mechanism behind this setup depends on the *physical carrier sense* with the understanding that every wireless workstation can listen to all other stations on the wireless network. However, it should be noted that every wireless workstation may not necessarily be able to hear all the other wireless workstations.

One solution to this problem is to define a second carrier sense method. The *virtual carrier sense* permits a wireless workstation to reserve the medium for a certain period of time using RTS/CTS frames.

For example, the first wireless workstation sends an RTS frame to the access point. The second wireless workstation will not hear the RTS. An RTS frame has a duration/ID field that designates the measure of time for which the medium is reserved for the next wireless transmission. This reservation information, used in the network allocation vendor (NAV) of all stations, is used to detect the RTS frame.

The access point answers the CTS frame when an RTS is received because it contains a duration/ID field that designates a measure of time for which the medium was reserved. When the second wireless workstation (stated above) does not detect a RTS, it will detect the CTS and update the NAV. Thus, collision is avoided through using hidden nodes from other wireless workstations.

RTS/CTS is utilized with respect to user-specified parameters such that it can always or never be used with packets that exceed a designated length. Note that DCF is the basic media access control method necessary for all wireless workstations. In addition, there is an optional extension to the DCF called the *point coordination function* (PCF) which yields the functionality for *time division duplexing* (TDD). TDM is the ability to deal with time-bounded and connection services.

Open System Security

The 802.11 standard provides security through two primary methods: authentication and encryption.

Authentication is the mechanism by which one wireless workstation is verified to have authorization to talk to a second wireless workstation within a specific WLAN area. Authentication is created between the access point and each wireless workstation when in "infrastructure mode."

Authentication has two specific modes: open system and shared key.

As described earlier, an open system allows any wireless workstation to request authentication, whereas the wireless workstation receiving the request may enable any authentication for any request. It may also enable access from only those wireless workstations on a user-defined list.

A shared system, however, only allows wireless workstations that have a secret encrypted key that can be authenticated. Note that shared-key authentication is only possible for systems that have an optional level of encryption functionality.

It's All About...Timing

Station clocks are synchronized at certain periods of transmission by a time stamp beacon. When working in infrastructure mode, your access point operates as the timing master and produces all the needed timing beacons. You can sustain synchronization to within 4 microseconds (give or take delay due to propagation). The timing beacons also function with respect to your power management. There are two power-saving modes pertinent to your needs: awake and doze (Figure 10.5).

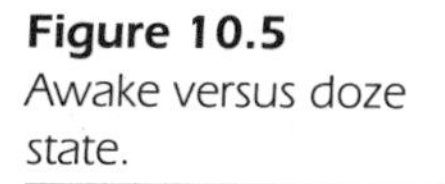

Figure 10.5 Awake versus doze state.

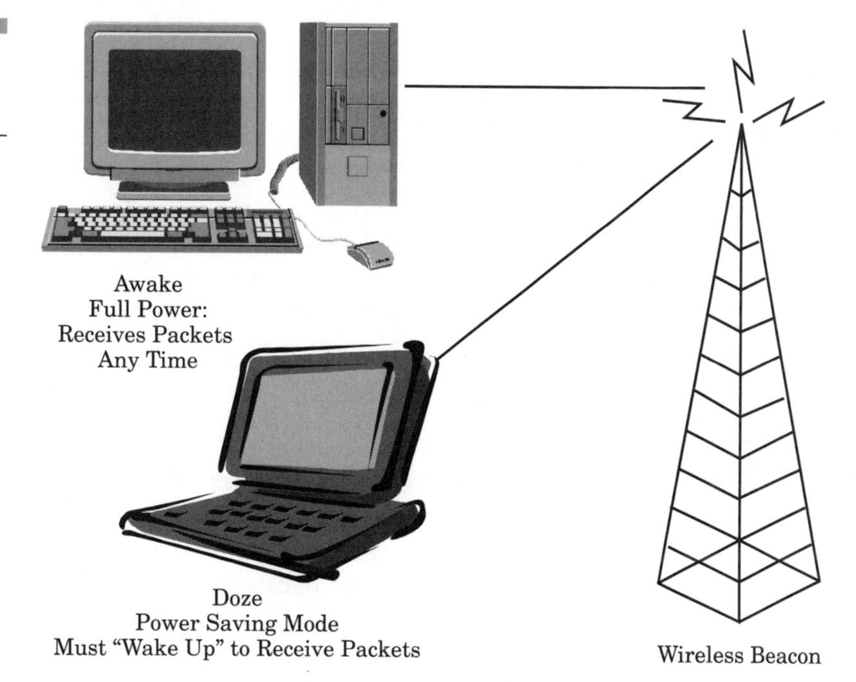

When working in "awake" mode, each station operates at full power so that it can receive packets any time. Each node must tell the access point of its intent prior to entering a "doze" state. In this mode, each node must wake up every so often so that it can monitor the network for beacons that inform the access point that there are messages for it waiting in the queue.

System Roaming

When dealing with open systems and frequency specifications, we must note that roaming plays an important role in identifying basic message format types. One of the elements necessary to support those areas of WLANs not covered by network vendors involves the interaccess point protocol (IAPP). IAPP enhances the interoperability of roaming wireless devices, regardless of manufacturer. IAPP addresses roaming that uses either a single extended service set (ESS) or roaming that occurs between two ESS units.

The problem with system roam is that it is far too easy for any computer in a Windows, Macintosh, or Linux environment to roam onto an 802.11 network. The hardware from each network vendor was designed for interoperability, so it is a simple matter for anyone on virtually any platform to roam easily (using an open system of authentication) and see all your network resources, intranet, file shares, and even access all of your network printers. The concept of system roaming is to be able to limit and define which users have access and what resources they can view. If you leave your wireless system completely open, it becomes a vulnerable target that anyone can exploit. Such vulnerabilities, using freely available spectrum communications, can leave your network open to attack and make it easy for someone to view, steal, modify, or even corrupt mission-critical data.

Conclusion: Spectrum Safety!

The purpose of this chapter is to provide a bit of insight into what people call the "inherent" security features of one method of spectrum communication over another. From experience, I have learned that there is no such thing as 100 percent security, nor is there a way for anything to be "inherently" more secure just because it uses a better protocol or means of transmitting information.

Many people might say that because FHSS hops from frequency to frequency, it is much harder (if not impossible) to figure out how to hack into it and eavesdrop on wireless workstations. The real truth is that if someone is intent upon listening to your traffic, breaking into your wireless network, or determining how to find out information—they can do it.

Your best defense is to find a way to make it harder for the hacker to break into your system. However, using FHSS rather than DSSS isn't going to be the way to protect your systems. You should never be lulled into a false sense of security because someone tells you that FHSS is more secure than DSSS. In truth, there is no real advantage to one method over another. In fact, when you consider these two choices you should remember that you can listen to a radio station on either AM or FM, and while the two methods of transmission are totally different—can't you still hear the station loud and clear on your radio? Even though these transmissions are on different bands and sent through different mechanisms, you can still hear them just the same. Think of FHSS and DSSS as you would AM and FM.

If you want true security on your system, remember to utilize encryption techniques and never use an open system of authentication (as discussed elsewhere in this book). If you leave your system open, you leave yourself vulnerable to attack.

Remember, no system is secure "out of the box" and don't ever believe anyone who tells you differently. Whatever the default values are for passwords, network protocols, and transmission standards—you should change all of these settings immediately. Hackers buy the same boxes and wireless equipment you do, they know all the default settings, and always use these items against unsuspecting users who have not taken the time to examine their systems and find out how to protect themselves against transparent intrusion attempts.

Hackers seamlessly enter your system and make very sure that they do whatever they need to so that they will not be detected. If you are careful and remember all the elements of security for each operating system, platform, and technology—you can realistically improve your chances so that you are not a victim of a hacker and therefore protect your systems against attack, intrusion, or any other form of unauthorized access!

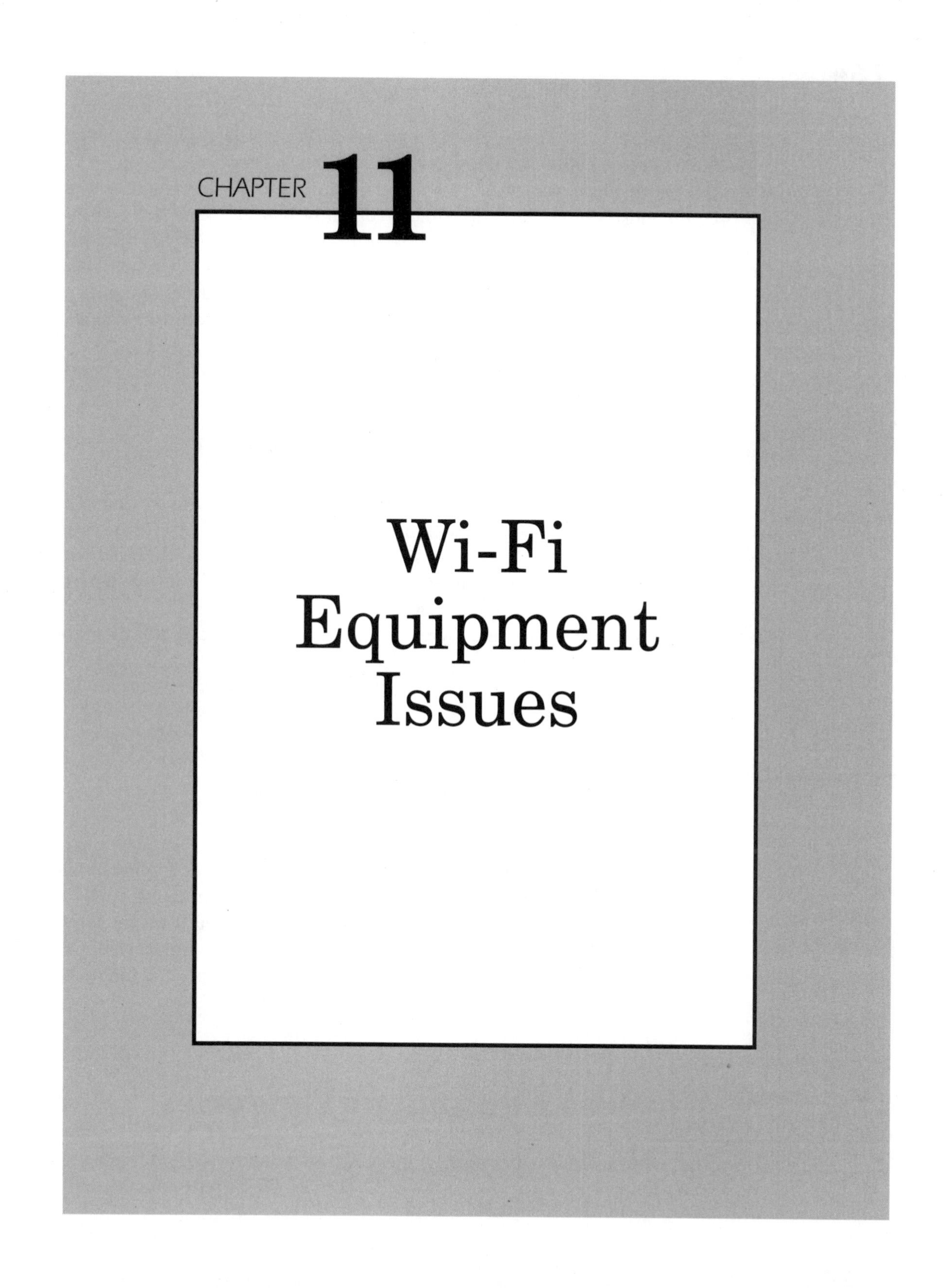

CHAPTER 11

Wi-Fi Equipment Issues

This chapter describes limitations in the criteria set by 802.11b to establish the reasonable use of low-level encryption to overcome present WEP limitations; however, 802.11b equipment purchased before the improved equipment is available will almost certainly have to be retired rather than upgraded. It seems probable that the upper-layer solution will require the services of a technician skilled in this area. Currently, the physical-layer issues making 802.11b vulnerable to a variety of denial of service attacks probably cannot be removed without substantial renovation of the existing DS PHY portion of the standard.

Issues in Wi-Fi Deployment

As convenient and useful as a wireless network is, there are a number of significant limitations that will affect the ways in which you can share your mission-critical data. Most WLANs work in tandem with wired networks in an effort to easily expand internal network coverage as departments grow or add new employees.

Wireless application deployment (Figure 11.1) uses the 802.11 wireless standards for items including:

- Collaborative code building
- Remote monitoring of the facilities
- Mobile access to database applications
- Mobile computer-based training (CBT) applications
- Wireless video distribution

The evolution of multimedia-rich applications requires higher bandwidth hardware for the wireless deployment of all the above applications. The current popular standard, 802.11b, is hindered by its speed barrier of 11 Mbps. This speed is really insufficient to run modern applications, thus initiating the need to move to the higher-speed standard of 802.11a, which supports speeds as high as 54 Mbps.

Wireless Equipment Vendors

The wireless LAN is becoming a staple in most businesses and will continue to grow as a core piece of the IT puzzle. Wi-Fi represents an effi-

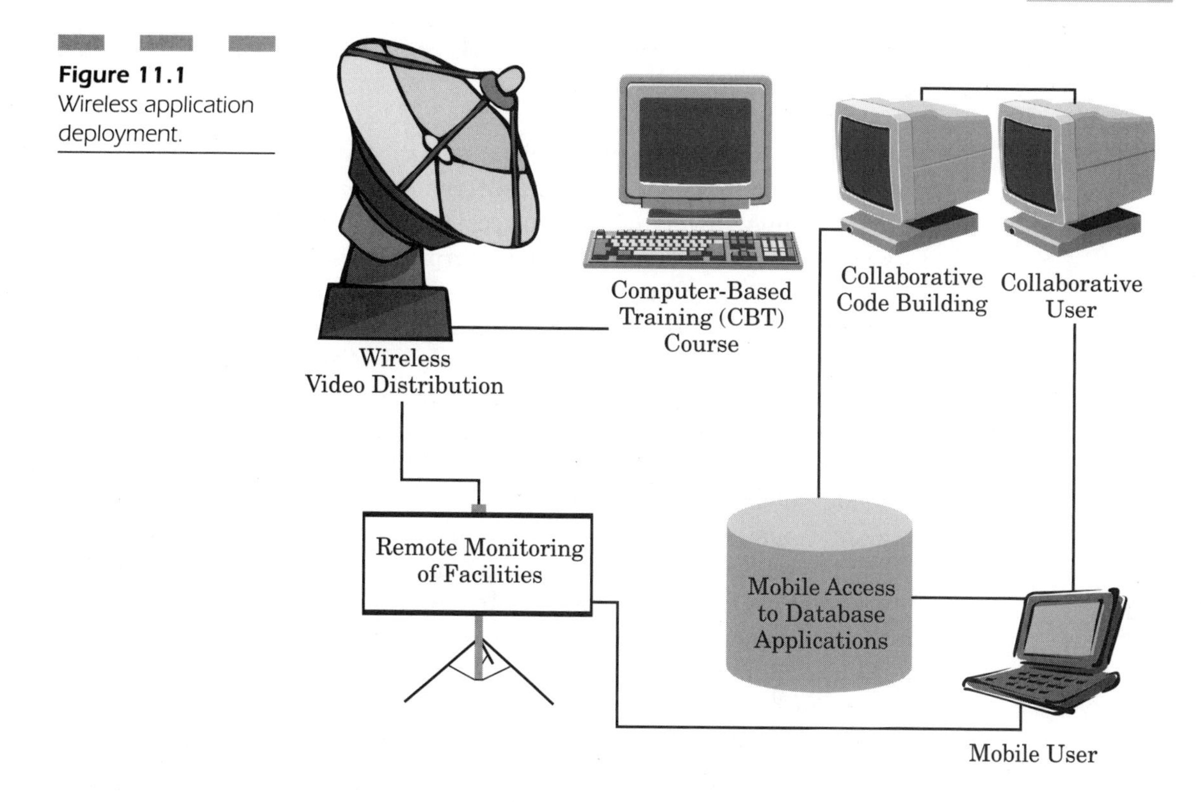

Figure 11.1 Wireless application deployment.

cient office solution for home and business users who require remote connectivity or the flexibility to access information from a mobile device regardless of location.

Wireless equipment vendors are focused on creating products that have broad enough compatibility to function seamlessly with different end-user requirements. These devices have the same benefits and disadvantages as any radio-based spread-spectrum technology. The majority of wireless computing devices now being built (including PDA-based telephone) incorporate either 802.11b or Bluetooth as standard built-in features that function in a variety of wireless networks.

WLAN Equipment Considerations

Wi-Fi equipment vendors are governed by several factors that dictate how the products we buy evolve with respect to our growing list of mobile applications. In order more fully to understand how these considerations apply, it is important to look at the factors pertinent for the equipment we purchase, such as:

- **Security**—In equipment development, security lacks both design and configuration. Wireless equipment vendors are working feverishly to design super levels of authentication and encryption schemes into wireless devices. The idea is to incorporate more secure methods for data transfers through the wireless equivalent privacy (WEP) encryption standard. Although WEP is easily defeated by any hacker with enough determination, using it (even with all its problems) is the least you can do to prevent hacker attacks. A newer generation (WEP 2) promises to incorporate more secure functionality, but the likelihood of its really being a hacker obstacle is low, since it is based on an easily defeated encryption mode. You will find most wireless equipment vendors support WEP, but the real truth is that you need to employ additional layers of protection, at a minimum, to safeguard your WLAN. One possible way to protect yourself is to incorporate IPsec virtual private network (VPN) functionality into your WLAN to achieve more powerful encryption, authentication, and key-management technologies. However, with each added level of security, you will degrade your overall performance unless you upgrade to a faster incarnation, such as from 802.11b to 802.11a.
- **Cost**—Deploying wireless networks is far more cost effective than laying wired LANs into new areas of your corporate facilities. Not only can you move freely throughout large areas of your production facilities, but you eliminate the cost of expanding your wireless infrastructure as your company grows. Wireless 802.11b devices are so comparable in price that it doesn't cost you significantly more to use a WLAN as opposed to a wired LAN. When you factor in the cost of installing and deploying a wired LAN against little or no cost of installing a WLAN, the wireless network ends up costing you less.
- **Network management**—So that you can manage your wireless network more efficiently, many wireless equipment vendors provide you with the ability to monitor and control the functionality of your wireless networking equipment remotely through easy Web-based interfaces. Using these settings, you have all the necessary inputs to ensure that you maintain proper network operation; however very few vendors of wireless LANs support advanced settings that allow you independent control of all the most important elements of your WLAN. Ultimately, your WLAN should support *simple network management protocol* (SNMP) so that you can manage TCP/IP internet-

work connections. SNMP is an effective tool that can be used to remotely monitor and control a wireless interface card's settings for both routing and radio frequency tables. SNMP remote links can be disabled and reset from the management console, and you can use this same functionality to monitor the performance statistics of all aspects of your WLAN.

- **Speed**—WLANs have broken through several speed barriers so that they can now compete sufficiently with wired Ethernetworks. The advantage that 802.11b users have is that they can upgrade to 802.11a, and in most cases the two types of equipment are compatible with one another. You can increase the wireless backbone speed of your WLAN, while ensuring compatibility with those users who still have integrated 802.11b network cards. Speeds from 11 Mbps using 802.11b are increasing to 54 Mbps using 802.11a. As this technology grows with the next generation of wireless equipment, these speeds will increase even more. As more and more multimedia-rich applications require more extensive bandwidth, increases in speed will be a logical step forward, while vendors maintain backward compatibility.
- **Interoperability**—One of the greatest advantages that 802.11 wireless networks have achieved is that, unlike previous generations of WLANs, there is a level of interoperability and compatibility between different vendors. If users wish to integrate 802.11b into a mobile device, it is still possible to have it work with the WLAN of their company or their homes even if the two points of the wireless transmission were created by different vendors. This level of compatibility has opened up an entirely new horizon that permits Wi-Fi users to roam across an entire wireless network without worrying about whether their equipment will function in different vendor environments.

Equipment Vendors

The 802.11 wireless standard went through a radical transformation in early 2000 as the standard increased its throughput from 2 to 11 Mbps. This made the wireless standard, at the very least, competitive with the old 10-Mbps wired Ethernetwork. It was at this point that vendors basically had to reformulate their offerings completely to ensure their equipment would function at the faster speed. The real trick was to make the pricing competitive in order to achieve mass adoption of this standard.

A number of vendors have taken advantage of this market, but several have established themselves as the main product vendors of 802.11b hardware. Most of these same vendors are working on 802.11a hardware that will also be compatible with 802.11b.

Wireless network equipment vendors The leading wireless network vendors include:

- 3Com
- Agere Systems
- AirConnect 11 Mbps
- Aironet 340 and 350 Series
- AirRunner
- BreezeCOM
- BreezeNET
- Cisco Systems
- Compaq
- Enterasys Networks
- Harmony
- Intermec Technologies
- Lucent
- MobileLAN
- Orinoco Wireless
- Proxim
- RoamAbout
- Zcomax
- Zoom Telephonics
- ZoomAir

Market Trends

When discussing trends in Wi-Fi equipment, it is important to detail how we got to this state in the marketplace. Standards such as HomeRF have become somewhat obsolete when compared to either 802.11 or Bluetooth. As much as Bluetooth has a good market presence for wireless devices, it is clear that 802.11 is holding on strongly to the market lead with built-in support in the major operating systems, including Windows XP, Mac OS X, and Lindows OS. Lindows OS is a UNIX-based operating system being developed in San Diego and has integrated support for 802.11. Lindows is a version of Linux that has the ability to run some Windows applications and function normally in a Windows-based networking environment.

Over the past decade, the primary market for WLANs has been corporate enterprise infrastructures which required that their users be able to use mobile applications throughout the geography of their production plants. Today, most residential and home office users have grown to need this capability just to do business in the tight confines of their working environment, where it is impractical to create a wired LAN infrastructure.

The most prevalent problem for home users is that fact that WLANs have such poor security. In most cases wireless equipment vendors configure access points and wireless network cards to function in an "open system" by default, without so much as warning the average user to change the default settings. Integrated security, which was supposed to be one of the strengths of wireless equipment, is actually one of the most easily exploited weaknesses.

WLAN vendors are now working on two elements:

1. Educating consumers to activate integrated encryption and change default "open system" settings so that hackers can't easily access their WLANs just by driving down the street with their 802.11b laptops set to "promiscuous mode," waiting to log into your private network.
2. Enhancing the encryption capabilities of their hardware. WEP is evolving a new (and theoretically) more powerful standard called WEP 2 that will enhance the security of wireless connections. Unfortunately, since WEP 2 is based on the flawed version of the original WEP, it is unlikely this capability will pose any obstacle for most determined hackers, who can force their way into accessing or eavesdropping on your wireless network.

Technology Issues

The concept of the WLAN has changed significantly in just the past few years alone. At first, wireless networks were used only to transfer small amounts of information from one department to another. There was no unified standard, so hardware was very slow and proprietary—there was no compatibility between different vendors.

Wi-Fi, now a standard most commonly represented in 802.11 networks, has increased in speed and in versatility. Information is now ported throughout large areas and can bring computing resources to the production areas of a company where no wireless infrastructure exists or can be implemented.

The technology for the WLAN is most commonly represented between a client and an access point, and there are a number of network setups possible.

Access Point-centric Configuration

The most commonly used wireless setup involves one access point and several 802.11a/b clients. When you install an access point, you enhance the range of your network. The access point functions as your wireless server, freeing up the resources of your wired server. The access point is connected to your wired Ethernetwork so that each client can access the network resources from every wired server as well as the file shares on other wireless/wired clients on your network segment.

Every access point can handle several clients, but this number is essentially restricted by two conditions:

1. How many simultaneous transmissions occur at any given time
2. How much bandwidth is consumed by a typical wireless transmission

Essentially, most applications today require much more bandwidth than their predecessors did. In a typical "real" situation, an access point can accommodate as many as 50 clients. However, as bandwidth demands increase, the number of clients the access point can host decreases proportionally with increased network resource demand. Multimedia applications are so common now that a typical 802.11b access point can only realistically support about 10 to 20 mobile devices at any given time.

Mobile Device Configuration

Wireless laptop users will require more bandwidth because they are running full Internet applications on a more diverse platform. Mobile PDA or PocketPC users will normally not require as much bandwidth, simply because of the limitations of these devices in supporting rich multimedia applications. However, while this may be the case for Palm-based applications, PocketPC requires significantly greater bandwidth. PocketPC computers have built-in multimedia players which consume as much bandwidth as any laptop computer. In addition, PocketPC supports "virtual desktop" applications so that you can use the full functionality of a PC. The richer your experience in connecting to your desktop computer, the more bandwidth your PocketPC will require. As mobile devices become more and more sophisticated, you can expect increased wireless network congestion for even simple tasks.

Peer-to-peer configuration In a peer-to-peer environment, an access point is no longer needed. In fact, all you need are two computers that have wireless network interface cards. These two computers form an independent peer-to-peer network so long as these two machines are within range of each other. This type of network is set up so that each computer only has access to the resources of the other machine; they do not have direct access to the server through a central type of access point.

Working with numerous access points In large networking environments, it is easy to see the limitations of using access points. As good as these devices are, they have only a limited range. Whenever you are trying to access network resources in a large production plant, it is far too easy to move out of range of your WLAN.

One way to improve your reception (in specified coverage areas) is to better position the access points in your company at greater heights so that you can achieve superior reception capabilities. This is only possible when you have a site survey to ensure that you have access points placed in areas where you can be certain you have adequate coverage without providing excessive coverage areas that hackers can easily exploit.

You can increase coverage areas by using multiple access points that have overlapping coverage so that you can maintain wireless LAN coverage throughout an entire area without moving out of signal range.

When you can allow your mobile clients to move easily from one access point to another, this is called *roaming*; it is a very useful wireless tool that enables you to provide seamless coverage completely transparent to the client in much the same way as a cell phone does. The main concern here, of course, is security. Increasing coverage places you at risk for eavesdropping or hackers' finding vulnerabilities in your network to gain access without your knowledge.

Building Extensions to Access Points

At times, it may be necessary to increase the coverage area for your WLAN. When you design your network, you can enhance your secure coverage areas by adding extension points that increase range to specific areas where employees will roam within the confines of your company grounds. The idea is to have each extension point extend wireless network range by relaying transmissions from one access point to another extension point. Extensions can be grouped together to send transmission packets from an access point to clients who are in other areas of your organization. The only problem is knowing where *not* to put an extension point within your organization. If you add extension points in areas that broadcast your signal into any public area, street, or residential area—you are literally inviting a hacker to take advantage of your enhanced signal strength to break into your network or eavesdrop on your confidential network transmissions.

Directional Broadcasting

One way to focus the transmissions of your wireless network so that you increase the range of your WLAN for employees but not for hackers is to implement directional antenna areas that enhance the coverage of your WLAN between corporate sites without sending your signal into public areas where they can compromise your system.

The antenna on one building within your corporate facilities is connected to its wired LAN, while the antenna of the adjacent building is connected to its wired LAN. You can point these two signal arrays at each other so that you maintain control of the radio frequency spectrum transmissions and focus your WLAN in areas where you need it, without sending your signal to areas you don't want to. In this way, you can maintain security while maintaining the functionality of your wireless LAN over moderate distances. You essentially save money without having to deploy more of your wired infrastructure.

Cost Concerns

Although WLAN equipment is initially more costly than wired LANs, you can actually save money when taking into consideration implementation costs for deploying LAN cables. There are several mission-critical factors you need to be aware of when you are contemplating costs (Figure 11.2) for your WLAN deployment:

- Manufacturer compatibility (with future standards like 802.11a)
- Manufacturer support

Figure 11.2
WLAN deployment costs.

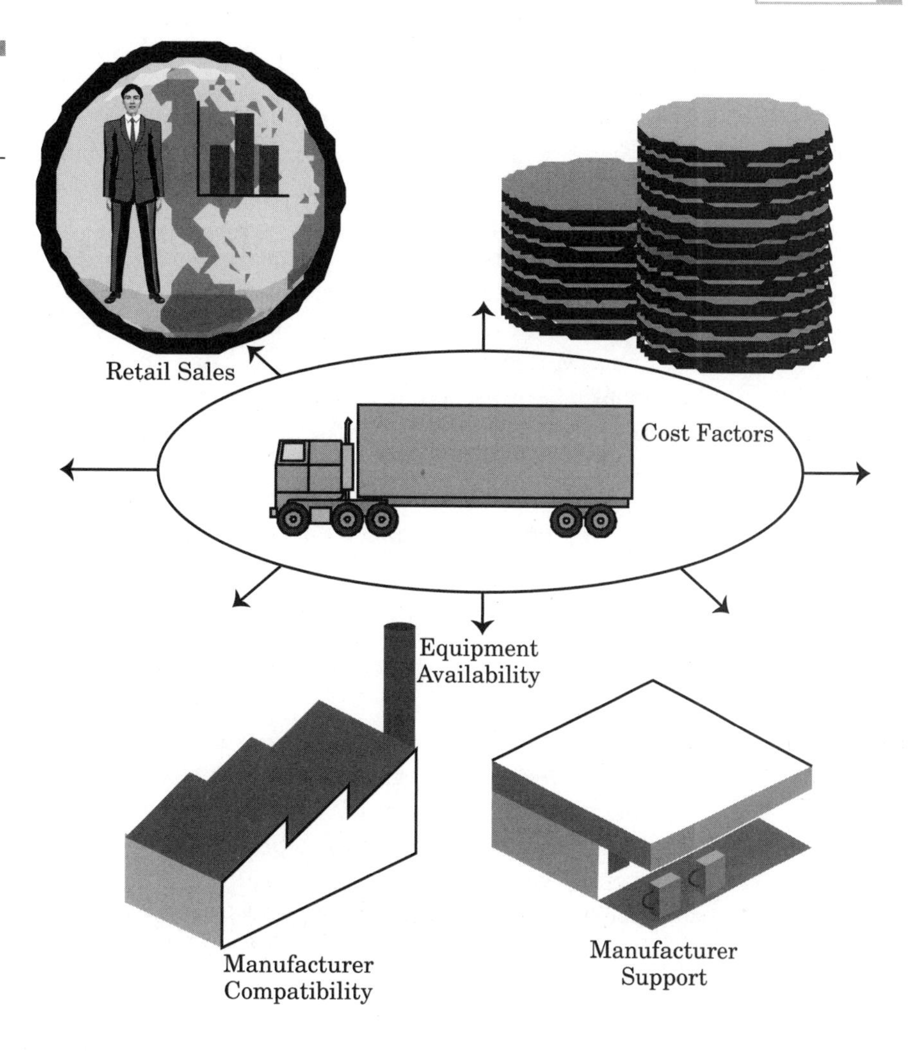

- Retail sales
- Cost factors
- Equipment availability

When you determine the actual cost for a typical WLAN, it is important to consider what types of computer devices you are going to use and how much each brand/version will actually cost you in the long run, taking into account operating expenses, software maintenance, and hardware upgrades. The factors that will determine the cost of your specific WLAN depend mostly on the devices described above, but you must also consider your monthly management, application development, and any outsourcing expenses you will incur in dealing with everything from installation to deployment.

In the majority of cases, your WLAN will actually pay for itself within a year, when you consider increases in efficiency and productivity. Equipment itself is not a limiting factor in today's market, due to ease of use and deployment.

The real issues involve the costs you will incur from a security breach. Without careful attention and a proper security vulnerability assessment, it is not uncommon for someone to hack into your network. Whatever cost savings you might have enjoyed with respect to deployment will be erased by analyzing the loss of data, business, and security. However, if you pay careful attention to your security needs up front, pay a little more for a proper security assessment, and maintain security guidelines for your system—you can realize the benefits of this technology and still save money in the process.

The Costs of Effective Security

Security has different types of costs that can both positively and negatively affect your organization. Security used always to be considered "negative" as most companies would say something like, "Oh no! We can't have a security audit, because that would make us appear as though we are not secure! Just the mere thought of mentioning security would make our customers think we are having problems!" However, the world has changed, even more so since 9/11. Security is no longer seen as a negative, but an essential positive that every company doing business *must have*!

Customers have come to expect that any company doing business on the Internet or with any type of wireless infrastructure must have certified themselves as secure. There is just too much personal information being transmitted over seemingly insecure channels on your network. Wireless networks have all the same flaws as wired networks, except that it is a well-known fact that most companies use neither the basic safeguards nor the proper levels of encryption to ensure that information is properly secure.

Wireless users are growing to represent an even greater number of company departments doing business. Normal LAN cabling is limited and can easily become damaged, forcing you to install new cabling at great cost. WLANs don't require you to maintain the physicality of your network infrastructure beyond the access point (server) and the mobile workstation (client). With so much personal information being transmitted wirelessly, it would literally bankrupt a business if it were to become public knowledge that hackers could sit in proximity to the server (Figure 11.3) and acquire items such as:

- Social security numbers
- Drivers' licenses
- Tax return forms
- Bank account numbers/statements
- Credit card numbers

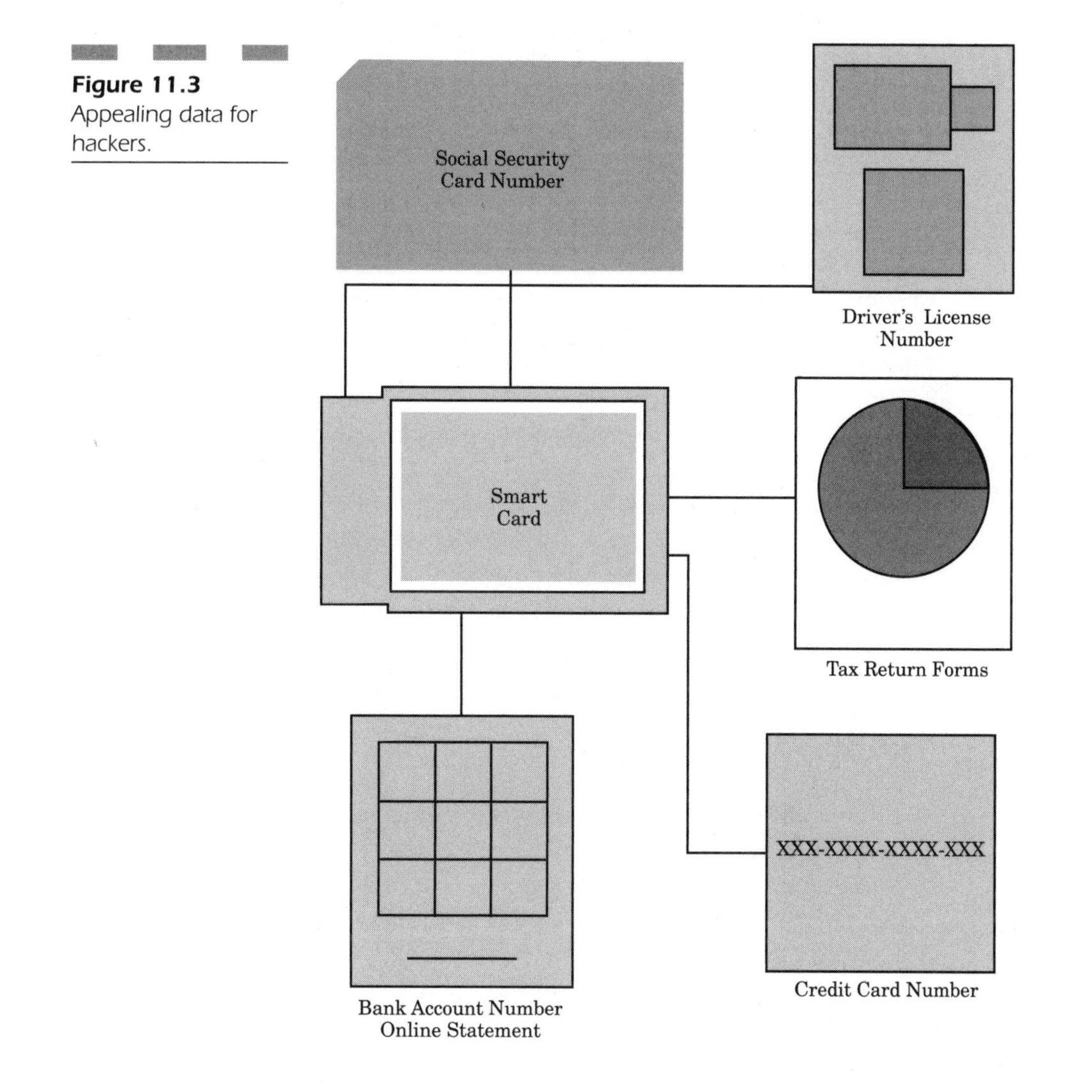

Figure 11.3 Appealing data for hackers.

The problem then becomes that you must convince your user base (employees and customers) that your wireless network has security comparable to that of your wired LAN. This means you must actually "prove" the concept of WEP, so that your wireless systems are as private and secure as your wired network. In order to accomplish this goal, it is important to draw an effective comparison between your wired and wireless worlds.

Wired versus Wireless Security

Wired LANs are much harder to compromise in terms of security because a hacker must physically connect to the network wiring in order to gain unauthorized access into the LAN. This means the hacker has to be inside the building or gain access through a public telephone line. You can implement greater physical security in your building, thereby preventing anyone who is not authorized from entering your facilities and accessing your LAN equipment.

The transmission signals from your WLAN are more fluid in that they are sent over the air from one building to another in your corporate environment. The problem is that the physical security mechanisms that were so useful in your wired networking environment are no longer useful in your wireless environment.

WLAN equipment vendors enable mobile networking cards to roam automatically until they access a wireless network. This means that security is no longer constrained to the actual LAN wiring, but instead is a problem for all the spaces in between buildings where the wireless signal is strong and can easily be picked up by any network card. In addition, wireless hackers can do a "drive by" to try and access your wireless signals from public streets.

Vendor Trials

Some companies originally placed severe bans on the use of any wireless LAN equipment because there was so much risk associated with using these devices. When WLAN equipment was first produced, there were so many problems that almost anyone could gain access; these devices become the least secure of any networking hardware. Most of these problems were due to malfunctions in WLAN vendor trials.

The result of these problems was that most retailers who were using Wi-Fi technology were forced to deactivate their WLANs because transaction data and credit card information were being stolen. Hackers only needed an active wireless device enabled just outside the perimeter of a retail store equipped with a WLAN. Users who were testing WLAN equipment outside these stores were intercepting confidential information!

The problem was the result of retailers who used point-of-sale data for both their pricing and inventory database programs. The information sent in these systems was *not* encrypted; it was easily intercepted by hackers who could then sell and distributed confidential information without fear of being discovered.

It is very important to note that WEP, once believed to be the wireless equivalent standard of privacy in a wired network, is now considered very insecure. Researchers at the University of California, Berkeley discovered a number of security vulnerabilities in algorithms upon which WEP was based.

The only way in which you can secure WEP is to use protocols (Figure 11.4) such as RADIUS, VPN, SSL, and IPSec.

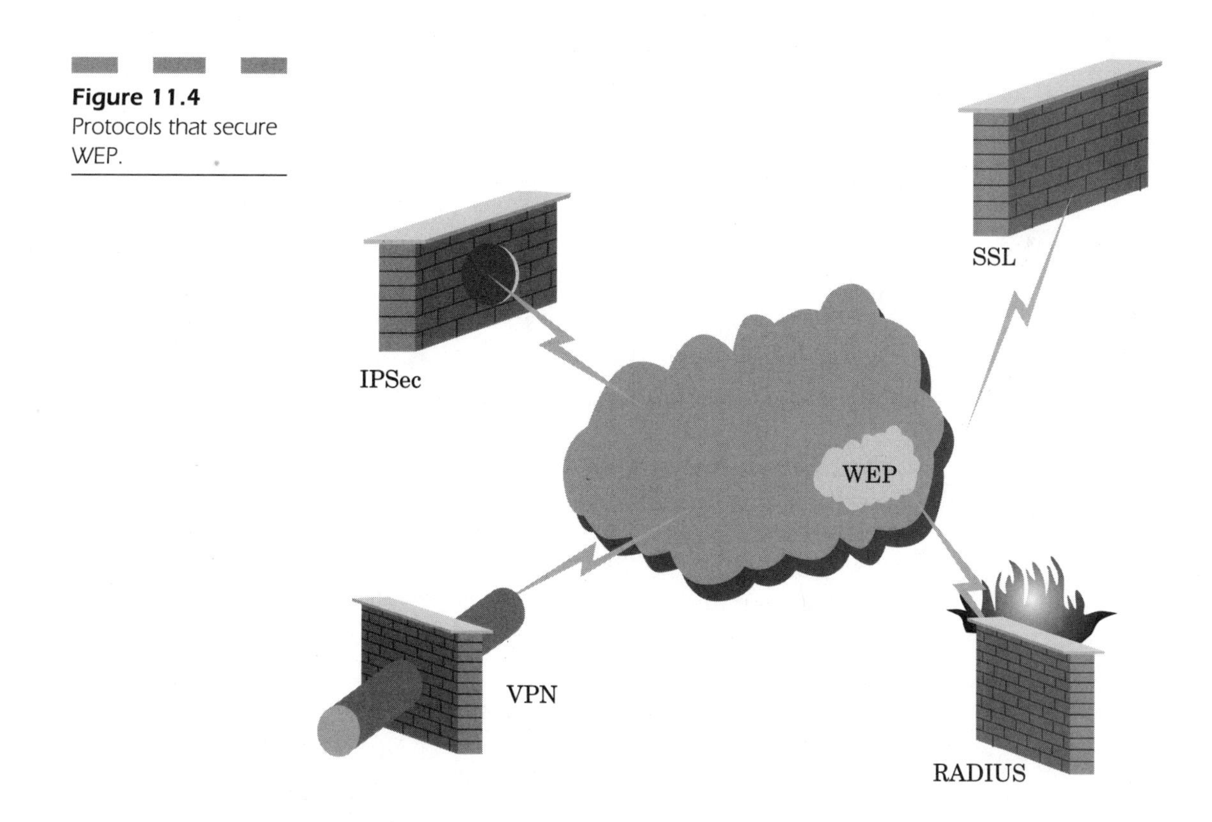

Figure 11.4 Protocols that secure WEP.

The idea is to add levels of security by having the software at each end of the wireless connection encrypt your data channels using its own specific algorithm. If you depend on the hardware WEP encryption built into your Wi-Fi equipment you are leaving yourself vulnerable to attack. But by using software level encryption, someone who does try to eavesdrop on your connection will not be able to make sense of your information sent in transit.

Conclusion: Next-generation Wireless Equipment

WLANs are becoming more than practical; they are becoming an ingredient essential to your communications needs. The combination of handheld PDAs and wireless connectivity is important for larger and more diverse companies to maintain information regarding inventory, eliminate errors, and increase overall efficiency through having information on demand.

Wireless devices are becoming important not only for corporate network usage but for numerous applications within the following industries (Figure 11.5):

- Healthcare
- Retail
- Manufacturing
- Hotel

In the healthcare setting, practitioners require constant access to handheld PDAs in order to retrieve patient, drug, and record information at a moment's notice.

Retail solutions, like those described in the previous sections of this chapter, give mobile point-of-sale terminals access to inventory, pricing, and sale information.

In the manufacturing industry, WLANs can deal with changing materials and information needs.

Even the hotel industry uses the WLAN to help employees stay connected so they have all the information they need to instantaneously serve the needs and desires of their customers in the most efficient way possible, without being tethered to a terminal. The problem is, of course, others in this environment can intercept the personal information of guests.

Established security really fails when it comes to adequate security, competitive standards (i.e., Bluetooth), deployment costs, network management functionality, system complexity, configuration issues, and finally the system interoperability between different wireless equipment vendors.

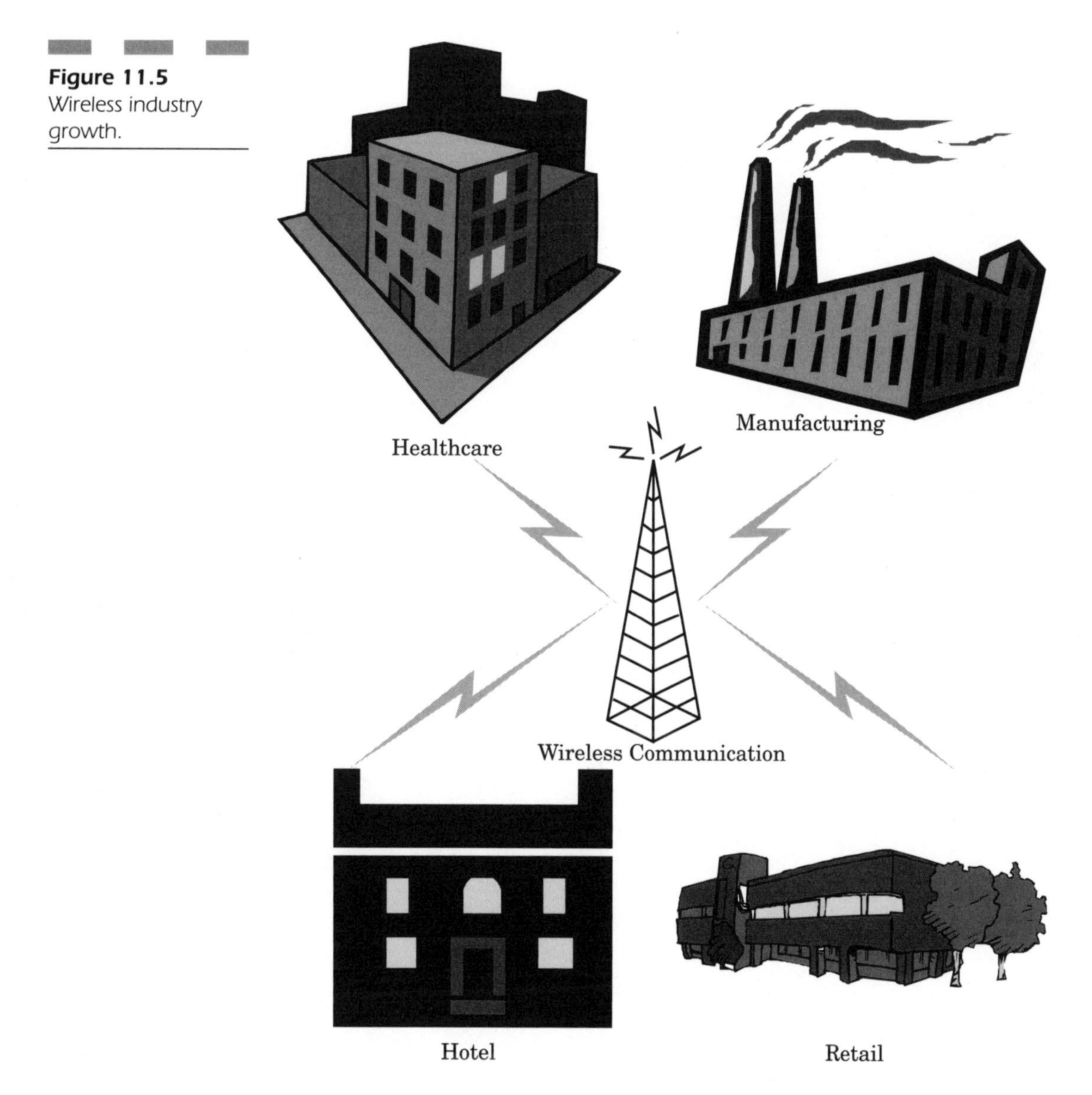

Figure 11.5 Wireless industry growth.

CHAPTER 12

Cross-Platform Wireless User Security

This chapter makes a detailed comparison of the wireless user security present in Windows, Macintosh, Palm, and PocketPC. It then describes a key weakness in any effective network security program—the users themselves. Studies consistently prove that the biggest security threat is from people inside an organization practicing poor security routines. Wireless LAN security, employing internal gateways or access points, is considered easier to control than the security of wireless handheld devices, which rely heavily on outside telecommunications companies.

WLAN Assignment Applications

The WLAN industry is growing by a factor of billions for everything from retail to healthcare. Among all these applications, fixed wired installations are not practical because of the nature of the job. Workers must be able to have constant access to their data sources over a specified controlled area.

In these applications, PC or Macintosh wireless laptops are the most powerful means of accessing information and having complete access to everything that the desktop can do. However, it is not always essential to have the power of a mobile desktop at your fingertips for most common information tasks. The laptop is still somewhat bulky and not always practical for most mobile tasks.

Enter the PocketPC and PDA. Microsoft PocketPC 2002 and Palm OS 4.x devices include support for Wi-Fi network access. You can access databases, connect to the Internet, retrieve e-mail, and do a number of data entry tasks on both these devices. These types of devices are far easier to transport than the laptop, their battery life is usually better, and they are easy to manipulate and use for a variety of tasks.

Cost Concerns

Both PC and Macintosh computer prices have dropped significantly over the past few years, although one cost that has remained the same involves the deployment of the actual wired infrastructure for a LAN. That is a fixed cost no matter what platform and operating system you use (Figure 12.1).

A great deal of hard labor is involved in buying the wiring needed for your LAN infrastructure, hiding the wire in floor or ceiling panels, and

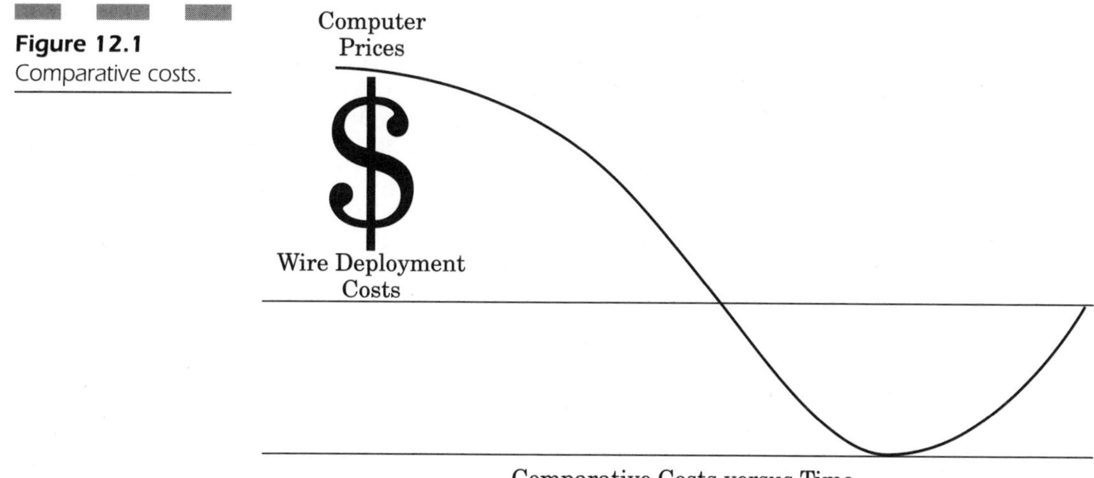

Figure 12.1
Comparative costs.

then deploying the proper connectors to each workstation. These connections last only so long, and eventually some of the wiring becomes defective, making it necessary for you to hire someone to restring another set of cabling.

As speed limits increased from 10 to 100 Mbps, most companies had to incur yet another expense to have new higher-capacity Ethernet cabling installed. Now, as we approach the need for greater throughput, gigabit Ethernet cabling will need to replace existing 100-Mbps cabling.

In wireless LANs, as speed throughput increases, it is not necessary to incur the same expenses. You need only replace one wireless router (access point) and its associated wireless network interface card. In theory, just having one wire break in the wired LAN is comparable in cost today to replacing a wireless network interface card, except the WLAN is far easier to maintain regardless of what platform OS you have.

Macintosh WLANs

Most Macintosh WLANs are built to be compatible with one another; however this is not the case for every product. Some products, even those built by the *same* vendor, may not interface with each other correctly. This is why, before the release of Mac OS X 10.2.x, third-party software was necessary in order to effectively and efficiently connect the Macintosh WLAN to the PC WLAN.

For example, in a typical corporate environment, you have Macintosh and IBM PC wireless workstation users. In most cases, you need the Macintosh users to be compatible with the wireless PC network. Note that as far as the Macintosh is concerned, an Airport card is just an 802.11b network card. The nomenclature might be different, but the fundamental concepts are exactly the same.

When working on an Airport-enabled Macintosh using OS 9.x, you need to install a program called DAVE, manufactured by Thursby Software. This program adds the ability for your Macintosh to connect or log on to a PC-based network simply and easily. Network file shares (or another computer) will simply pop up on your Mac desktop as a network drive. You can then read or write to the specified device through your wireless Airport connection (as your access privileges dictate).

If you are upgraded to Macintosh OS X 10.0.x or 10.1.x, you have two choices. One choice is to use the Samba interface command, through the Connect to Server option menu, manually enter the IP address of the file server you want to connect to (just as for Mac OS 9), and bring the file share of the PC network onto your computer.

Your second choice is to buy the DAVE product for Mac OS X and install the ease of having a hierarchical menu displayed from your Connect to Server option so that you can choose the file share you need without manually finding and inputting the specific IP address of the PC computer you need to connect to.

Both of these options are useful and allow your wireless Airport card to connect seamlessly to the PC. However, if you just upgraded or are planning to upgrade to Jaguar, the new Mac OS X 10.2.x, then you are in for a surprise because DAVE 3.x no longer works. In fact, you won't even be able to connect to your PC file server if DAVE 3.x was on your computer before you upgraded to Mac OS X 10.2.x. Thursby software does provide a fix that removes the program that must be input into the command line interface of the Mac. Mac users will have to use DAVE 4.x in order to regain the functionality of this nice program in Jaguar.

It should be noted, however, that Mac OS X 10.2.x includes its own version of PC compatibility software in its Connect to Server menu option that allows you to select a file share dynamically from a drop-down list for a PC. Mac OS X 10.2.x is a better operating system because its ability to connect to either a Macintosh, PC, or Linux network is all built in by default. In fact, I run a wireless PC network, and my Macintosh running Mac OS X 10.2.x connects better, faster, and more seamlessly to any wireless network fileshare I come into contact with—better even than my PC!

Lindows OS

Lindows OS, not to be confused with Microsoft in any way, shape, or form, is an interesting operating system offering. It is essentially a standard Linux platform with the ability to execute certain Windows applications. It is somewhat similar to WINE.

WINE is an implementation of the Windows Win32 and Win16 APIs that runs on top of X and UNIX. It is considered to be more of a Windows compatibility layer. It provides both a development toolkit (Winelib) for porting Windows sources to UNIX and a program loader that permits many unmodified Windows executables to run under Intel UNIX platforms. WINE also works with Linux, but Lindows is a more complex version that integrates Windows applications seamlessly. Some Windows applications work very well, other applications don't work at all. It depends on what is supported by Lindows and what is not.

Network connectivity in Lindows combines all the benefits of accessing UNIX and Windows networks. Lindows is no exception; it can access almost any file resource that any Windows computer can.

Wireless network connectivity is integrated directly into Lindows as part of its core operating system offering. Most Linux distributions (such as Red Hat and SuSE) have built-in support for 802.11b, though enabling them is not a simple task. On the other hand, Lindows has a nice built-in GUI that is easily configured through its network settings (in much the same way that Windows XP has integrated support) to set your SSID and other wireless settings.

Lindows, unlike Microsoft, does not support every possible hardware vendor. This means that whenever you search out a wireless networking card, it must support either UNIX or Linux in order for it to work on Lindows OS. Even though Lindows OS is touted as a replacement for Windows with "broadband" wireless network connectivity, at its core it is still Linux; you will have to find hardware that has Linux, not Windows, drivers, in order for it to work on your systems.

Orinoco Wireless

Orinoco Wireless produces a wireless NIC that works on almost every platform. The reason I mention this particular card is that this is one of the few companies to offer support for Windows, Macintosh, Linux, Novell, and Windows CE.

This card is an excellent solution because it is one of the few that can be natively supported in Windows XP, Red Hat Linux, SuSE Linux, and Lindows OS, to name a few. I have even seen this card working very nicely on Windows CE handheld devices.

Orinoco offers integrated encryption for the cards in two flavors:

1. Orinoco Silver Card offers 64-bit encryption and 802.11b connectivity
2. Orinoco Gold Card offers 128-bit encryption and 802.11b connectivity

This is one good example of a vendor who designed hardware to work on nearly every platform with seamless connectivity, no matter what computer or operating system you run (Figure 12.2).

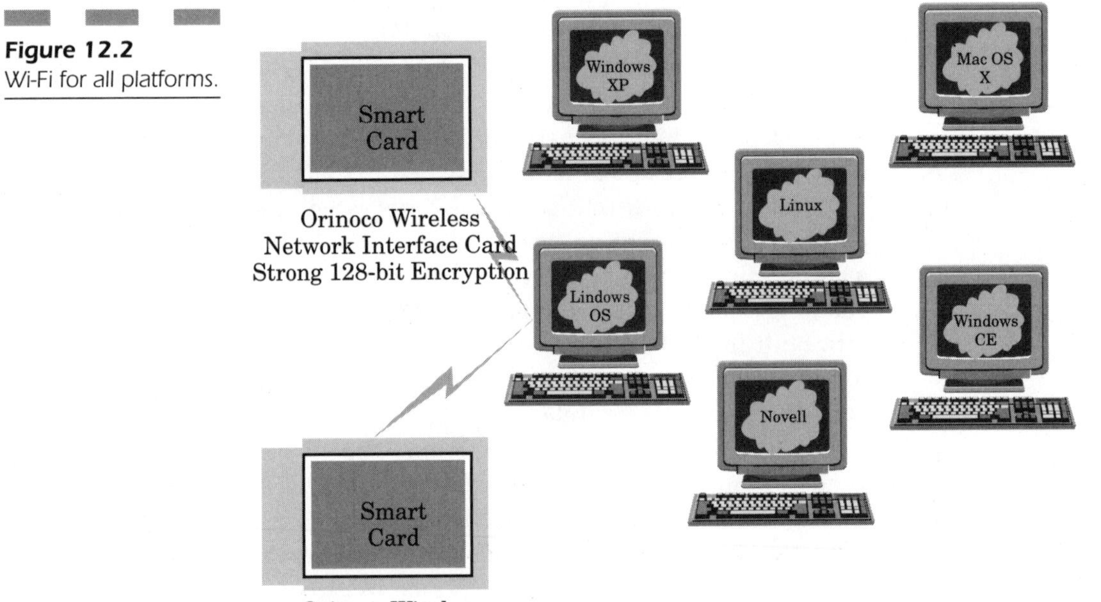

Figure 12.2
Wi-Fi for all platforms.

Handheld Devices

The security ramifications of handheld devices are both good and bad. The good side of these devices is that they are light, easy to carry, and have just enough processing power to enable you to access your information needs wirelessly from your corporate network.

Handheld devices such as Windows CE, PocketPC, and Palm-based devices don't usually have any file share that a hacker could compromise. Since there is no hard drive on these devices, there is a very low likelihood that they could accept any incoming data connections that would compromise the integrity of your data or constitute a security breach. These devices really have only enough memory to help you execute the tasks you need, because most of the work is done on the remote server to which you are connecting.

The bad elements of these types of devices are that they are too light and too easy to steal. They are usually small enough to fit into a shirt pocket, so it is very easy for the user to lay the device down somewhere and expect to pick it up later. Unfortunately, since most users configure these devices to retain all network passwords and settings in their memory, it is a very simple matter for a hacker to use these default settings and instantly gain wireless access to your entire corporate intranet with a stolen handheld device.

Knowledgeable hackers understand the file system on both Microsoft and Palm devices. There are a number of "hot sync" programs that can literally pull the entire memory, configuration, network connections, and passwords out of the device's memory so that they can be used to mount an attack against your system.

The new Palm-based devices come standard with integrated 802.11b connectivity. As storage and memory increase on these devices, it becomes more and more probable that secret and confidential passwords and identity information can be accessed by hackers who come into possession of these devices if they are stolen.

These wireless devices have the same WEP encryption capabilities as standard wireless workstations. Setting up security involves different menu options, depending on whether you are using a Palm or PocketPC. It is extremely important that you activate the highest level of encryption possible for these networked devices so that transmitted information cannot be intercepted.

Cross-platform Wireless Security Concerns

Wireless networking support was added as a core offering within the Windows XP Operating System. As 802.11 has seen enormous growth in many wireless network deployments, it is actually the lack of a WEP key management protocol that causes the primary limitation in its security, especially with respect to building a secure wireless infrastructure using access points as an interface to your wired LAN.

When you use manually configured shared keys, they often remain in place for extended periods of time. The longer they remain, the greater the chance that hackers can employ specific attack patterns to acquire your keys and decipher your network traffic. Security can also be compromised when you lack both authentication and encryption services, as this affects your wireless operations when an ad hoc or peer-to-peer wireless network uses wireless collaboration tools. This tends to explain why it is so crucial to have both authentication and encryption in your WLAN. Access control is one of the more important elements of security that incorporates the key management protocol within the 802.11 specification.

Initialization Vector Collisions

Since there have been noted problems with WEP, security concerns deal explicitly with items like initialization vector (IV) collisions. The problem lies with how the RC4 IV is employed to create the keys used to drive a pseudorandom number generator used to encrypt wireless network traffic. For example, the IV in WEP is defined as 24 bits, really only a small space that can be misused by reusing keys. Furthermore, WEP doesn't define how the IV is designated, so that many wireless NIC cards reset these values to a null value and then increment by one for every use. This means that once a hacker has caught two packets using the same IV or key, it is possible to discover information about the original transmitted packets of information.

Key Reuse

Key reuse constitutes a problem because keys can be compromised and used as a form of attack against your wireless system. These types of attacks do require about 6 million packets in order to determine the WEP key in a reasonably short amount of time. When stations all use the same shared key, the chances of increasing IV collisions goes up significantly, resulting in degrading the security of your network because the WEP keys are not changed often enough.

Evil Packets

When the hacker knows the actual structure of your encrypted packet, such as the header field, he can send "evil" packets into your network to change commands, spoof addresses, and perform many other tasks. Encrypted packets have an integrity check to make certain they have not been altered, but the integrity check within WEP can be changed so that it will actually be "valid" for the "evil" packet to be accepted by the receiver. When the hacker knows the location of the receiver, the address can be modified to reflect an unknown packet; thus the new destination can now be controlled by the hacker. If the packet is transmitted on a wireless network, then the access point will actually decrypt the packet and send it along to its false destination.

Real-time Decryption

Due to the small size of the IV in combination with the long-term key reuse that is so prevalent today, hackers can easily create a table of both IVs and key streams, adding to this table for every single packet that is decrypted. Ultimately, this table will possess all the possible IVs and can then be used to decrypt all your wireless network traffic in real time.

802.11 Security Issues

The most prevalent security issues having serious implications for cross-platform wireless computing involve key problems that have universal significance.

There is no per-packet authentication method per se that allows you to determine the source of a specific packet coming into your system. This leaves 802.11 vulnerable to "disassociation attacks" that force users to disconnect from the WLAN at any given time.

802.11 has neither a specified method of user identification nor of authentication. Without any central method of authentication, authorization, or accounting support, 802.11 is vulnerable to so many attacks that it leaves the system completely vulnerable.

Even when the RC4 encryption cipher is used, it is highly vulnerable to known attacks because there is no security or verification mechanism in play for 802.11 users. Making this problem worse is that some

WLANs set their WEP keys from existing passwords; this makes the passwords vulnerable if the keys are also determined.

Even with extended authentication, there isn't any support offered. Other security mechanisms vulnerable (Figure 12.3) include:

- Smart cards
- Certificates
- Token cards
- Passwords (one-time expiry)
- Biometrics

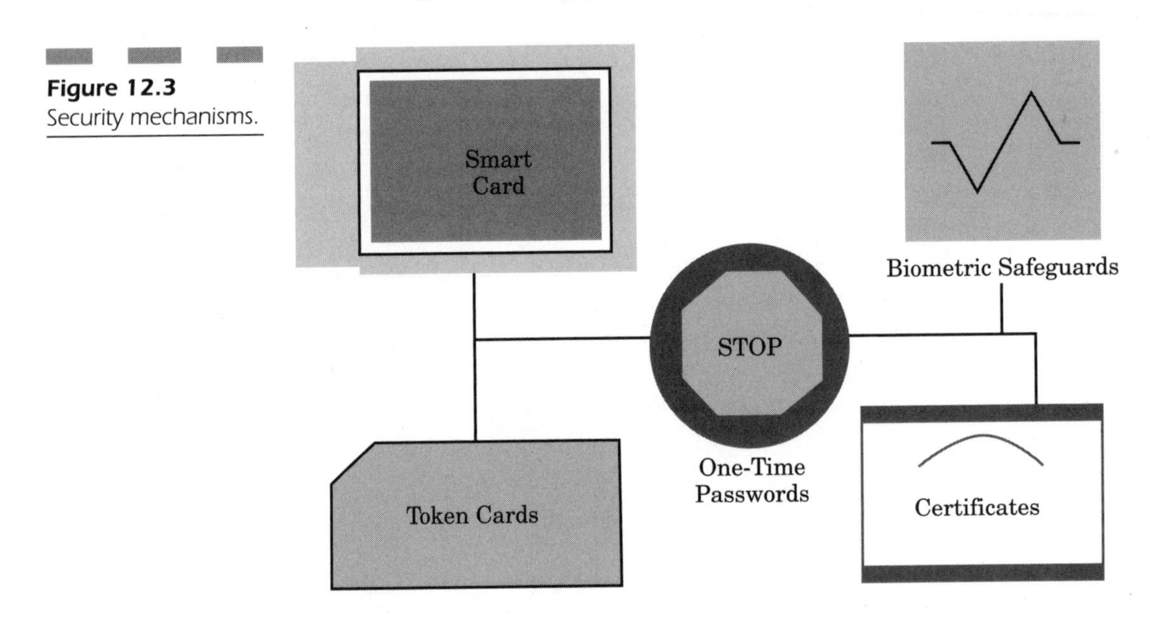

Figure 12.3 Security mechanisms.

Key-management issues (Figure 12.4) include:

- Rekeying global keys
- No dynamic per-station or session key management

The market hype for most 802.11 products is that they offer security that is essentially "equivalent" to that of wired Ethernetworks. The truth is that wireless networks are vulnerable to attack. If you are unaware of all the problems that exist for a typical WLAN, it makes you that much more vulnerable to compromising your internal network infrastructure for anyone with enough time and tools at their disposal.

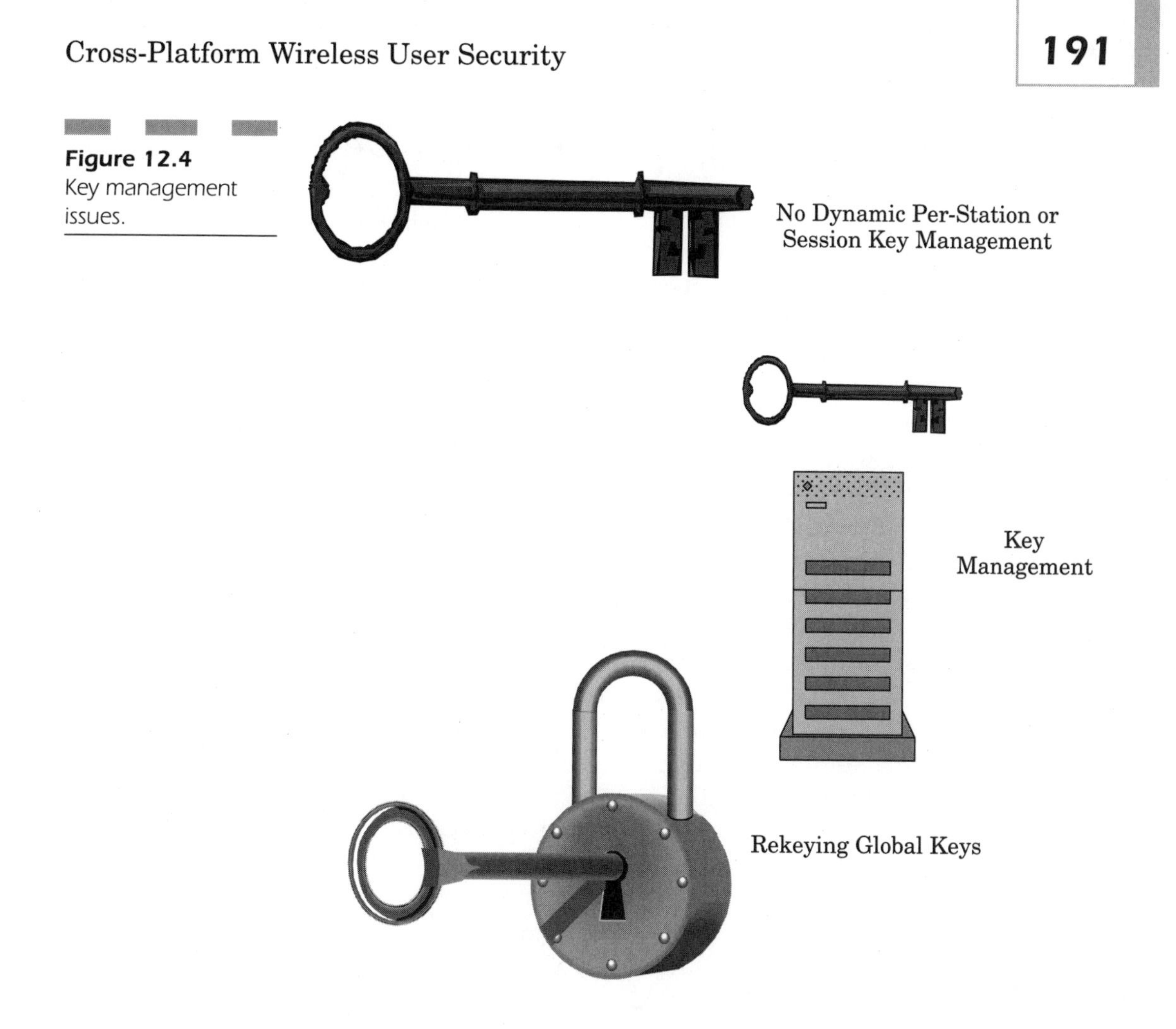

Figure 12.4 Key management issues.

By taking the proper precautions, you can effectively learn how to protect yourself, based on these types of vulnerabilities. Understanding how to establish password policies, add the highest level of encryption possible, and screen out MAC addresses from wireless NIC cards that don't belong on your network are just some of the ways you can protect yourself.

Even if you are only able to slow down a hacker from accessing your network, that might give your administrator enough time to see if security is being violated on any of your computing platforms, identify the problems, and correct them before your WLAN has difficulties.

Windows XP Wireless Connectivity

In an effort to provide its latest version of Windows with the ability to deal seamlessly with wireless networking capabilities, Microsoft has worked with a number of companies that are within the IEEE standards groups to define a "port-based network access control" standard that better defines 802.11-based wireless networks.

The fundamental understanding behind 802.11 is that it does not require the same WEP keys to be used by all of its stations because it allows a station to maintain two distinct sets of shared keys: a per-station unicast session key and a multicast/global key.

Existing 802.11 deployments currently support only "shared" multicast/global keys; however, this will undoubtedly change by 2005 to support per-station unicast session keys.

Managing these types of keys is often very difficult; the current 802.11 security for access control does not scale well for either large network infrastructures or ad hoc networks. Furthermore, there is no definable interaccess point protocol (IAPP), which makes it very difficult to manage keys when stations actively roam from one access point to another. Without IAPP, authentication must restart upon each new connection.

802.11 is a standard for network access control dealing with each specific port. It is used to offer authenticated network access to users. When dealing with network access control for specific ports, you need to employ the specific physical parameters of the switched LAN itself in order to offer a method of authenticating devices on your LAN. It is also very useful as a means of preventing access to a given port whenever you are unable to adequately authenticate a wireless user attempting to use your WLAN.

Windows XP WEP Authentication

When you have several access points set up to use the same WEP key, each will implement added optimization routines so that the wireless NIC will try to execute 802.11 authentication using the WEP key received from the original access point as the *shared key*. Once this routine is successful, the access point will instantaneously add that station to its authenticated list of stations. However, if the authentication routine fails, the NIC will open up to authenticate that access point and finish its 802.11 authentication.

The access point needs to be able to determine if a station that has openly been authenticated to the access point can effectively complete the 802.11 authentication. This means that the access point must be able to determine whether a station is "open authenticated" or executed with "shared-key authentication," as shown in Figure 12.5.

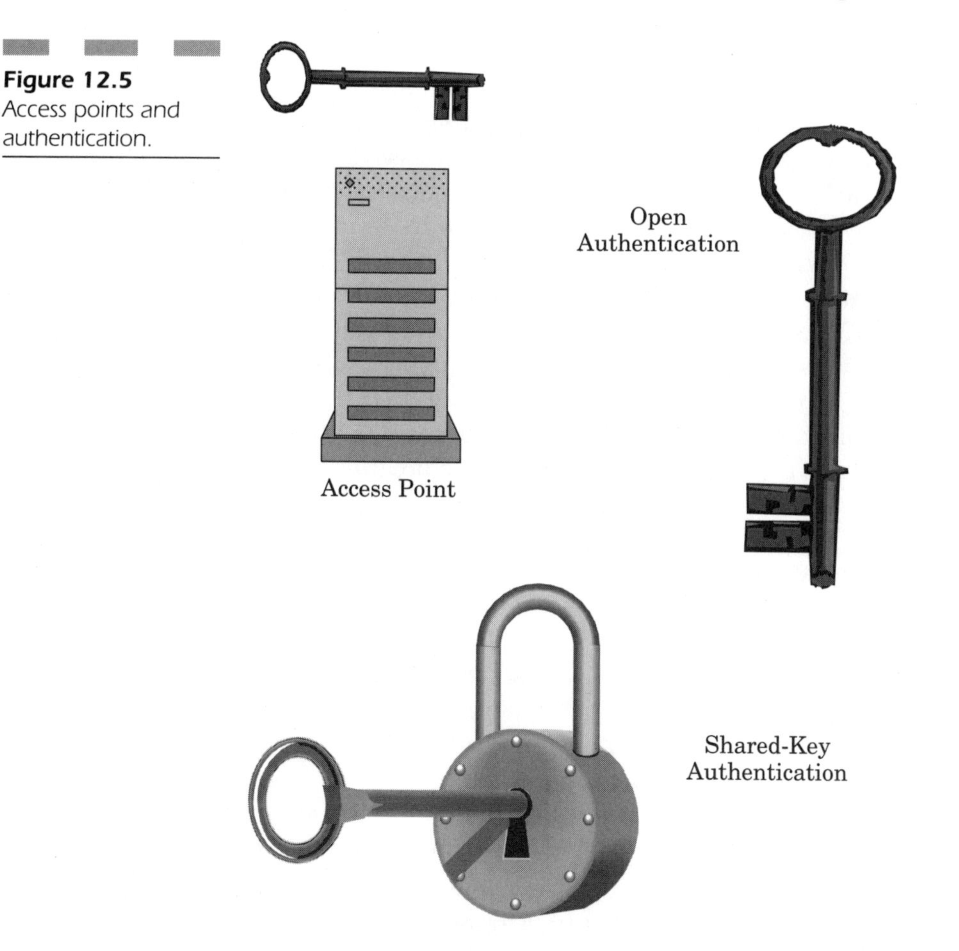

Figure 12.5 Access points and authentication.

When the station does acquire access to the new access point using "shared key authentication," then the 802.11 authentication routine is still started by the new access point so that it can update its record-keeping system. When you initiate wireless station network connectivity through shared-key authentication, the new access point will start 802.11 to make certain that the wireless station does not experience any interruption in its network connectivity. When the wireless station is not able to complete the 802.11 authentication successfully with the new access point, then the wireless network connectivity to the wireless station through the controlled access point port will be killed in order to make certain that you retain the highest possible level of network security.

Windows XP Wireless Functionality

Windows XP has a variety of features and functionality when it comes to supporting 802.11. For example, it supports automatic network detection and association. This is useful for wireless NICs, since the operating system can tell them to use a logical algorithm to detect what wireless networks are available, and it can associate them with the most appropriate connections.

Media sense is a Windows XP function that can be used to decide when a wireless LAN NIC can roam from one access point to another. You can also determine whether you need to reauthenticate or alter your specific wireless configuration.

Windows XP also supports network location functionality, which permits applications to be notified whenever a computer is roaming through the WLAN. This information allows programs to update their individual network settings based automatically on a given network location.

Wireless NICs support power mode changes and are summarily notified whenever the power is coming from a fixed A/C adapter or a battery. This feature makes it possible to conserve energy when you need to.

The goal of adding all these features is to give Microsoft the ability to implement a secure wireless solution and make certain that network traffic is confidential.

There are also a number of vendors who create add-on 802.11 security solutions that are locked into their technology or hardware infrastructure. When solutions are proprietary, it becomes enormously difficult to determine how capable these technologies are for protecting you from known attack patterns. The majority of password-only solutions are often highly vulnerable to a hacker type of "dictionary attack," making these types of vendor solutions highly insecure; they often cause more havoc than they prevent.

WLAN NIC Vendors

To better understand how vendors provide cross-platform solutions, we can study the types of solutions that are offered today.

Proxim and Cisco represent a large number of WLAN products that are very well known for healthcare and manufacturing solutions. These products represent nearly a third of the WLAN products on the market.

Agere is a company spun off from Lucent Technologies to produce an Orinoco wireless network solution that is very popular because it supports Windows, Macintosh, Windows CE, Novell, and Linux. The universal nature of this card has made it attractive, and it is one of the leading 802.11 wireless network hardware vendors because of its nature.

Table 12.1 lists those wireless network card vendors who have a predominant share of the WLAN market.

TABLE 12.1 Primary Wirless Network Card Vendors

Vendor	Operating System	Maximum Throughput
Buffalo AirStation	Windows 95, NT, and 2000	11 Mbps
Cisco Aironet	Windows 95, 98, NT, 2000, and CE	11 Mbps
Orinoco Wireless (Agere)	Windows OS (all), Windows CE, Macintosh OS, Novell, and Linux.	11 Mbps
Proxim Harmony	Windows 95, NT, and 2000	11 Mbps
Symbol Spectrum	Windows 95, 98, NT, and CE	11 Mbps

Conclusion: All Vendors Must Get Along!

Wireless networking has truly become part of the corporate infrastructure. In almost every application or business unit, wireless networking has become integral to what we do and how we work.

The biggest problem involves security and how we must maintain a level of heightened security despite the obvious flaws and problems with 802.11 and its deployment. Realizing that WLANs will never truly have privacy factors equivalent to those of any wired network, knowing how each hardware vendor can use simple features that enable encryption, screen out unwanted stations, and support security in any form are more protection than having nothing!

Even spammers have taken advantage of the flaws prevalent in wireless networking. Hackers now "drive by" unprotected WLANs in the parking lots of many companies, use their internal SMTP outgoing mail server, send thousands of spam e-mails (clogging up mail servers), and then simply drive away.

Most companies have WLAN equipment serving a variety of different hardware and software platforms, but the universal factor in all these implementations is that you can enable security that restricts who can easily access your wireless network. You can minimally prevent users from breaching your network by keeping a log of every machine that has access to your system. If you know who will access your system, what computer OS will use your WLAN, and the unique MAC identifier addresses of each wireless network interface card that will log into your network—then you have the basic tools to protect your cross-platform WLAN and prevent security breaches from destroying the validity and functionality of your wireless networking infrastructure.

CHAPTER 13

Security Breach Vulnerabilities

Wired networks have always been advantageous in that they are fast, stable, and provide a hard-lined connection to the network. Unfortunately, wired networks restrict the user's mobility throughout the office. In addition, it is very difficult to deploy new users on a wired LAN, especially if your location is geographically diversified from one department to the next.

Wireless LANs provide users with the ability to work virtually anywhere. The user can change his location at any time while keeping his network connection. There are numerous cost and time advantages in connecting new users on the network without having to extend your network infrastructure.

Mobile users have the most to gain from this technology because they can access their e-mail through their cell phone, PDA, or any other mobile device.

The greatest weakness of wireless technology is that it uses over-the-air infrared and radio transmissions to send data between two network points, and these transmissions can be easily interrupted or intercepted.

Intercepting Wireless Network Traffic

Wireless LANs running either 802.11b or Bluetooth have the advantage of being able to work from virtually anywhere; however, both these technologies depend on open communication from point to point.

Ad hoc networks represent two different computers connecting by one wireless link; however, it is more common to use "access points," routers that have the capability to route LAN traffic on both wired and wireless networks. Access points can serve hundreds of different computers. The greatest vulnerability of these types of devices is that mobile devices enter a type of "promiscuous mode" where they will search their local area for any access point that can host a connection into the local area network. When the mobile device finds an access point, it immediately locks onto that device and forms a transparent and seamless connection into the LAN. Unless these access point devices are configured with some level of security, virtually *anyone* can connect!

Wireless 802.11b

Wireless 802.11b is the most commonly used wireless LAN protocol. This type of technology uses direct sequence spread spectrum (DSSS) to produce bitstreams that are sent in the 2.45-GHz ISM band. Speed varies for this technology depending on how good your signal is, but can be as high as 11 Mbps or as low as 2 Mbps from as far as 1000 feet.

The main concern with 802.11b is that it is sometimes far too easy to attack its vulnerabilities. The next section will discuss the most common attacks that hackers can use to try and gain access to intranet and network resources.

Proximity Attack

Information is literally bursting out of some wireless networks. Many people equip their laptops with 802.11b cards and attach small antennas only a few centimeters long to the external ports of the wireless network cards to give the signal some gain. With this type of setup, the hacker can walk or drive outside a building that houses an 802.11b network, set his card to promiscuous mode, and then from the street pick up the signal and access the wireless network without anyone in the company even realizing it!

Many department stores set up wireless cameras to transmit digital images of different sections within the store to a main computer to monitor everything going on in the store. These types of cameras are easy to set up because they have no wires to install. Unfortunately, the same type of proximity attack used to access a wireless LAN can be used to intercept the video feeds of these wireless cameras. Potential thieves can literally case the routine of the store and its workers to devise a method for stealing from the store without anyone knowing.

When a hacker tries to sign onto the network, his first step is to try to determine your service set identifier (SSID), which corresponds to the name of your wireless network. The hacker can then use that SSID to access your wireless LAN by having your router assign him an address through DHCP. In most cases, however, it is not even necessary for the hacker to know your SSID to gain access and get a dynamically assigned IP address. Most wireless routers are so "user friendly" that whenever a mobile device has a blank entry for the SSID it is to lock onto, it will look for any SSID in range of the device and roam right onto the LAN.

You can impose specific restrictions on your network by assigning wireless users a predefined media access control (MAC) address (which essentially is a unique number that identifies your network card) such that only machines you want to have access can gain entry into your LAN. But, like most technology items today, that too can be easily spoofed. Mobile devices now have the ability to copy the MAC address and use the number they copy as their own. For all intents and purposes, it is always possible to gain access into your wireless LAN if there is enough time, motivation, and desire to get there.

The MAC address is a hardware address that uniquely identifies each node of a network. In 802.11 networks, the data link control (DLC) layer of the OSI reference model is divided into two sublayers:

1. The logical link control (LLC) layer
2. Media access control (MAC) layer

The MAC layer interfaces directly with the network media. Consequently, each different type of network media requires a different MAC layer (Figure 13.1).

Figure 13.1 MAC and LLC layer representation.

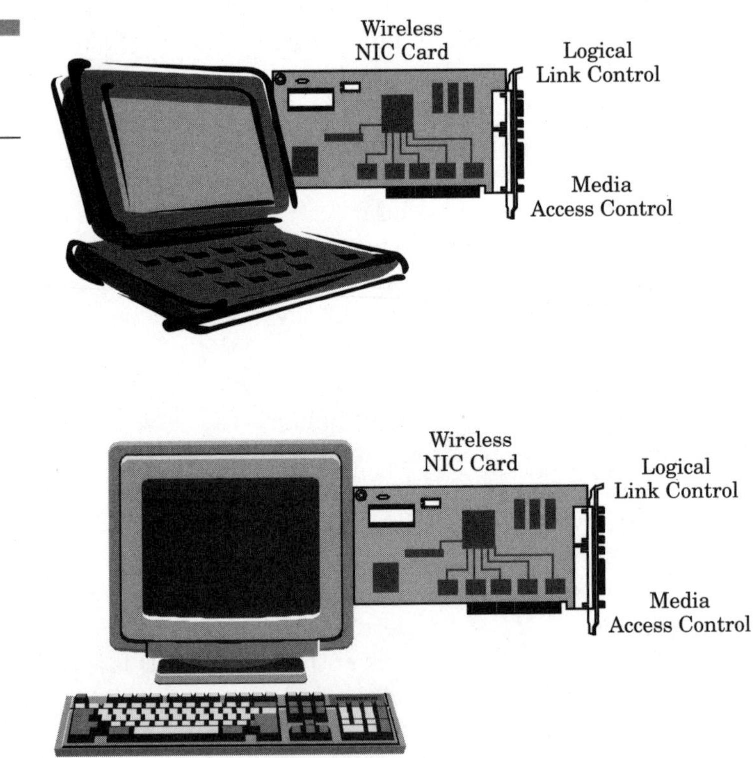

Securing Your Network

There are some simple but effective measures you can take to prevent someone from accessing your network without your permission. Most of the wireless routers on the market allow you to configure an internal firewall to keep any open port on your machine from being used against you in an attack.

Another simple way is to prevent the "leakage" of any radio waves from your building and onto the street or parking lot where someone can pick up your signal. You can impose a specific level of shielding around the walls of the room that houses your wireless router, thereby restricting the signal strength to the immediate area within your building. This would make it harder for hackers to roam onto your network, but you may have discontented employees who no longer have as good reception on your wireless LAN as they would have liked.

WAP Attack!

Wireless phones have become so popular that they have largely merged with personal digital assistant (PDA) devices. Now, a mobile employee has essentially a mini-computer that can be used to access e-mail, Web pages, and even information from your corporate database anywhere and at any time. While this may sound like a good thing for productivity, it leaves gaping holes in your security for people to take advantage of.

These devices are vulnerable to attacks that break through wireless transport layer security (WTLS), which is essentially the same as SSL/TLS in the TCP/IP protocol. WTLS is used to protect the data transmitted between cellular or wireless devices to the WAP gateway.

There are several types of attacks that occur in the wireless domain. These include trying to crack specific plaintext data recovery, datagram truncation, message forgery, and key-search shortcut attacks. These types of attacks exist because of bad protocol design and poor implementation. These attacks succeed largely because of people who understand the methods behind cryptography and how the WTLS protocol works.

Encryption

When you think of your wireless network, you must consider several options. It is an incredible convenience to be able to establish a LAN wirelessly, but remember that specific trivial measures should be take to secure your network.

When you set up your wireless router you can set an encryption key which would keep an unauthorized person from trying to log onto your network. There are three levels of encryption possible, depending on the hardware you purchased (Figure 13.2).

- Off—No data encryption
- 64-bit encryption
- 128-bit encryption

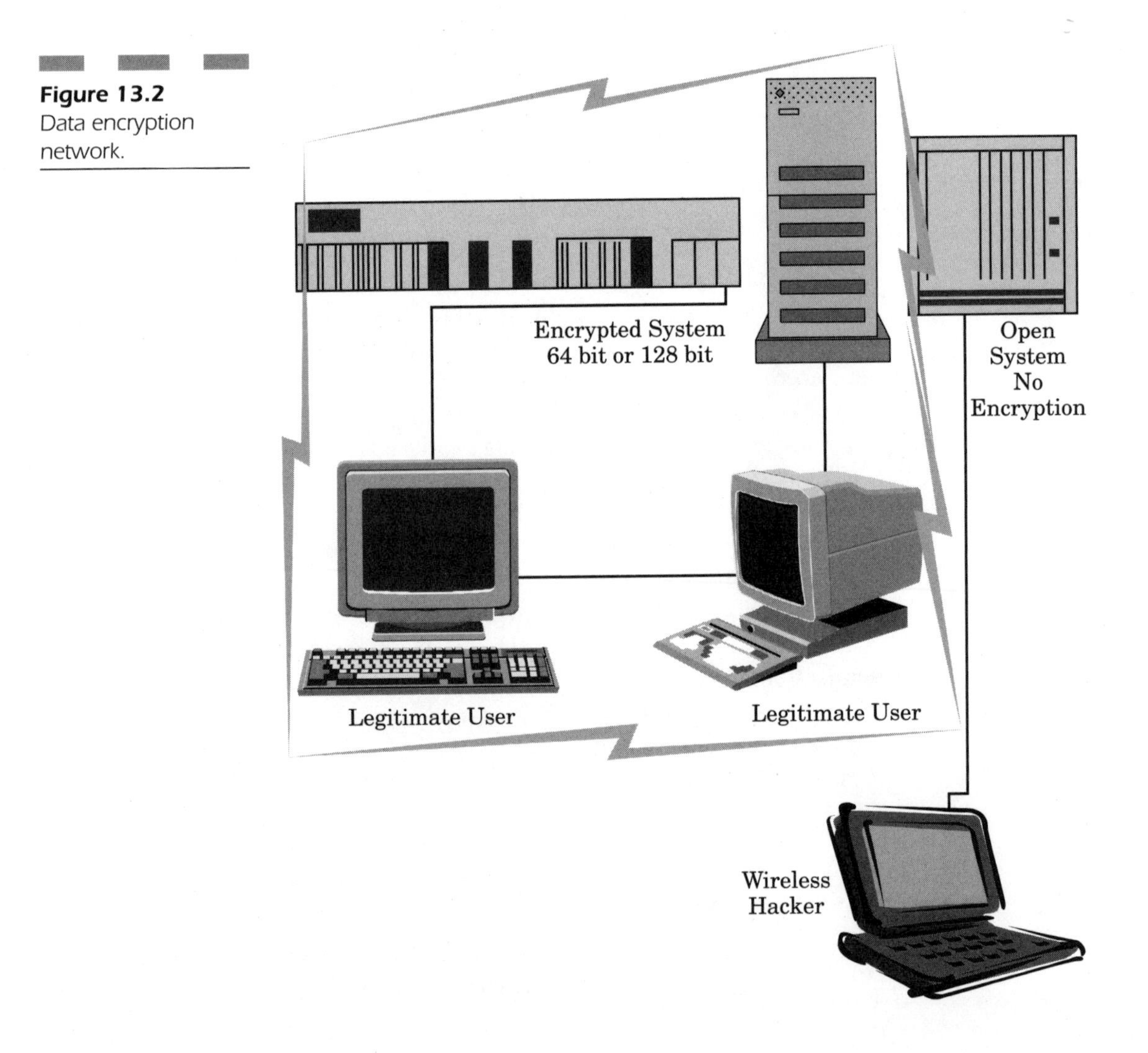

Figure 13.2
Data encryption network.

By default, all wireless routers are configured with *no* data encryption whatsoever. Most users and administrators don't even realize that by setting a very simple parameter in the Web configuration dialog, you can easily establish an encryption key for at least the 64-bit encryption. Using this simple encryption method, if anyone tries to eavesdrop on the wireless traffic in your network, he would have a difficult time decoding your session.

For a few more dollars, you can invest in network cards that support 128-bit encryption. This provides you with a much more comprehensive level of protection that makes it even more difficult for a hacker to try to decode any network session between a user on your wireless network and the wireless LAN access point.

You can establish an encryption key known only to the access point and to the user; this makes it very difficult for many people to roam onto your network.

It is important to note that while encryption protocols stop the vast majority of hackers from roaming onto your network or decoding your network traffic sessions, it is not impossible for any wireless encryption key or scheme to be broken. In fact, given enough time (as little as a few days or weeks), a hacker can determine even the key to your 128-bit encryption scheme and roam onto your network. He can then, theoretically, decode your network traffic session and see all the data transmitted across your network.

Commonsense Measures

Make certain you take all necessary measures to ensure that only those people authorized to access your network have true access during regular business hours. You can look to the log activity to determine if there is an overload of network access that is consuming all your bandwidth. If you keep an eye out for suspicious activity, you can usually make certain, by taking all the precautions you can, that your systems run properly.

PnP Networked Devices

It is a common misconception in many companies that any security vulnerabilities present in their WLANs will not necessarily affect the wired

LAN to any great extent and cause damage or corruption of data. The actual truth is that you *must* to consider your WLAN as an extension of your wired LAN. For all intents and purposes, every resource accessible on your wired network is accessible on your wireless network.

The most overlooked resources available on your LAN are plug and play (PnP) network devices. The best example of this technology is the networked printers within your organization. These devices have their own built-in network servers and their own individual IP addresses. These devices have no protection whatsoever and anyone (with very little effort) can find these devices on your network to use or to destroy their functionality.

Malicious attacks against your network can be as simple as taking over all the printers on your network; launching this type of an attack over a wireless LAN is even easier. A hacker who cannot readily log onto your network to access file shares or other computer resources can very easily scan your network for network printers. These printers show up readily on the "Network Neighborhood" of any Windows computer, but are even *easier* to find on a Macintosh!

Windows Users

Windows users can click on their "Network Neighborhood" icon and see everything in their local LAN segment. The Windows machine sees an 802.11b network just the same as if it were connected to a wired LAN. A user who clicks on the global icon to see the entire networked area can see all the different LAN segments within the entire organization. It is simple to expand the hierarchy tree for every computer in each LAN segment to see which computers are connected to printers. Many LAN printers show up as a printer icon with computers attached. In most cases, these printers are freely accessible by anyone who can see them on the network. Why not? What possible harm can a hacker do a freely accessible network printer? You will see the answer in the next section that details the vulnerability of these devices. Adding any one of these printers to your computer is easy; sometimes you can simply double click the printer icon to add the necessary drivers to your system. You can start printing to these networked printer devices instantaneously and no one would know you have added this capability to your workstation until you start printing to any device.

Macintosh Computers

Macintosh computers running OS 8.x through 9.2.x have an application called the "Chooser" that is part of the operating system. This application not only gives the user access to every local printer, but also to programs like AppleTalk and DAVE. AppleTalk is Apple's built-in method of network printing. This program allows you to do a search on your local area or wireless network for any printers that support the AppleTalk protocol. Many of the mainstream network printers (like those produced by HP or Xerox) support this protocol.

The program DAVE, by Thursby Software, also gives the Macintosh both the freedom and flexibility of Windows computers have. The Mac user can easily allow DAVE to show him every single computer workstation and printer accessible on the network. Only now, the Mac user has the ability to add and use not only printers accessible on the Windows network, but on the Macintosh network as well. In many cases, you can set the settings on the Windows network differently from those on the Mac network. While you may need special instructions to add a printer via TCP/IP printing, (or the PC equivalent networking protocol NetBIOS) the Macintosh could easily add this formerly inaccessible machine via AppleTalk. This is why having a different platform like the Macintosh available on your 802.11b network can open up several new types of vulnerabilities you were not aware of before.

When using a Macintosh with OS X, you have an entirely new set of optional features that you can use to find networked printers. As far as the Macintosh is concerned, any 802.11b network card is the same. Not only can you use AppleTalk to connect to printers, but you can search for and add any TCP/IP printer on your network too.

The DAVE program (useful for all of the Macintosh operating systems described here) helps the Apple user connect to any network share or printer when connected to either a wired or wireless network. Any protocol that a Windows machine can use, a Mac can use more efficiently, without very many complicated settings to maintain.

Linux Boxes

PCs configured with Linux have the same type of flexibility as Mac OS X or Windows for adding network printers. Linux can operate seamlessly with an 802.11b card. Not only can it access all the same network printers, workstations, and resources as the other machines, it can also utilize networking tools not available to the other platforms to break into portions of your wired LAN through the WLAN without your knowledge. There are a growing number of hacker tools for the Linux platform and these tools are expanding in power and breadth every single day. Linux computers offer the versatility of seeing your wired LAN through its wireless NIC card more easily than on other machines. It can initiate hacker attacks, denial of service (DoS), as well as a number of other attacks.

Needless to say, adding a plug and play networked printer is quite easy. There are a few major versions of the printing utility in Linux that allow these devices to emulate Windows (through the SAMBA equivalent networking protocol), NetWare, TCP/IP, and a host of other protocols too. These machines can gain access to devices through networking protocols (i.e., NetWare) that you may not have known existed on your networked printer device. The fact that Linux is so versatile opens up an entirely new set of vulnerabilities for your entire network.

Hacking the Network Printer

We have just defined the security vulnerabilities that networked printers can experience from Windows, Macintosh, and Linux platforms. Now we can define such vulnerability. Figure 13.3 shows how a network printer can be just as accessible on your WLAN as any other networked computing device.

Adding these devices is simple. For any of the computers we described, there is no difference whatsoever in a computer connected to a wired LAN as opposed to one connected to a wireless LAN. They all have the same power and connectivity. The difference is that a computer on a wireless LAN need not be anywhere inside your building. A computer on your WLAN can be a PocketPC that nobody can detect or a Linux computer sitting in a car in the parking lot just outside your building, but still within range of accessing your wireless access point.

Printers are not normally configured with safeguards and have open connectivity right out of the box. But if a hacker wanted to use your wireless network against you, he could easily connect to all of your printers during off hours on the WLAN (without having even to step foot inside your building) and cause them all to print garbage data until you exhausted your complete paper and ink supply. Can you imagine coming into your office the next day and seeing your floors filled with reams and reams of paper from a hacker sending print jobs to them all night long!

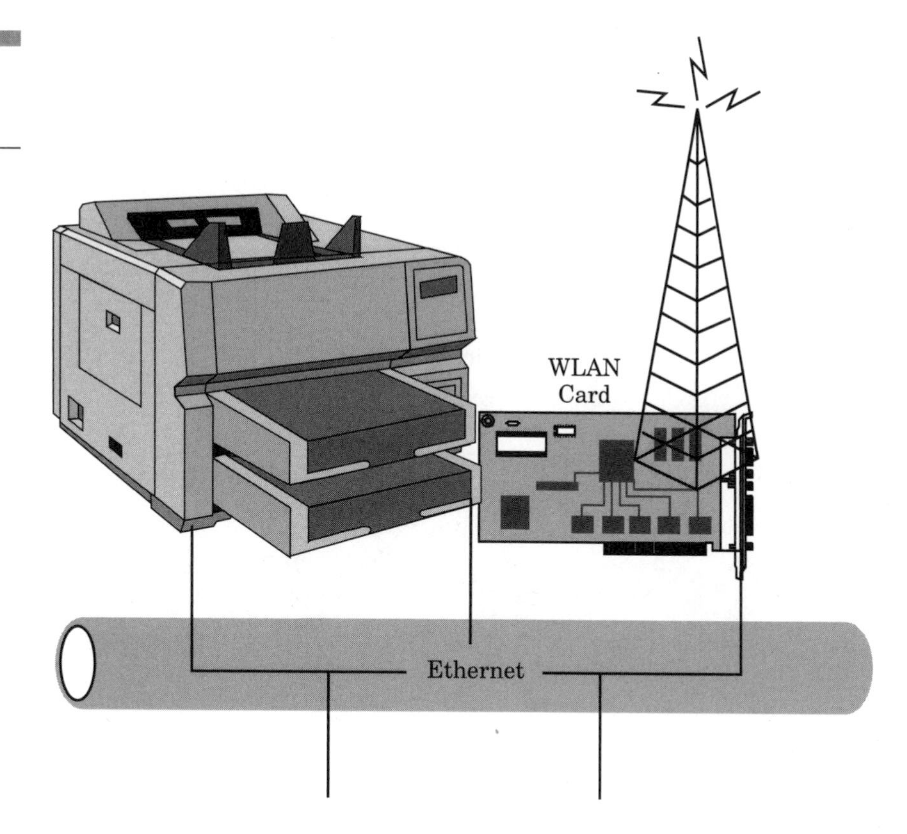

Figure 13.3 Wireless networked printer.

If such an attack happened during the day, a hacker could easily send rather large graphic documents to each printer. This transmission of large files would consume all your bandwidth as the files traveled from your wireless LAN to the networked printers hooked up to your wired LAN. Nobody could use the Internet, do file transfers, or even have enough bandwidth to read e-mail! The congestion of such an attack would not only destroy the functionality of these very expensive printers, but tie up your network so that you couldn't even perform the simplest business activities.

Printer Servers

Many of the printers released today have built-in Web servers that allow for easy remote configuration from virtually anywhere within the network. They are advantageous configurable entities, but present a risk of unauthorized users who gain access to this device.

A hacker could use an internal WLAN to gain access to the printer's Web server and reconfigure the machine so that it won't print for any user anymore. This can be a catastrophic event, as these machines are configured out of the box to allow anyone in the network to change the settings.

These Web servers also have a configurable option for "Security" that enables you to defend yourself against unauthorized configuration changes. There are configurable settings for:

- Login/password administrative access
- Authorization settings for various features and functionality

The administrator can input both login and password settings to restrict access to the printer's configuration dialog on the network. This would require that someone know these private bits of information before any changes could be made, so that even if a hacker does break into the network and access the Web server, it would be very difficult for him to effect any changes in the printer's configuration.

There are also several authorization settings that an administrator can set on the printer to block specific features from people on the internal wired or wireless network. Administration, printing, and firmware/software upgrades are by default accessible to any user on the network. However, these settings can be changed so that only an administrator can access e-mail reports, printing utilities, and software maintenance. This protects your networked printer devices against unauthorized use. The key is to know how to configure these settings properly; otherwise, by default, you are wide open to an attack by anyone on any wireless platform.

Defending Against Attacks

A good defense is having a knowledgeable offense. This means that you must look at every computing device, printing device, and networked storage resource device on your network as a potential way in which hackers can breach your system and gain access to important resources (Figure 13.4).

Each device on your network needs to be examined in terms of security including:

- Networked printers
- Networked attached storage (NAS) boxes
- Wireless routers
- 802.11b servers
- Web servers
- File servers
- Network fax servers

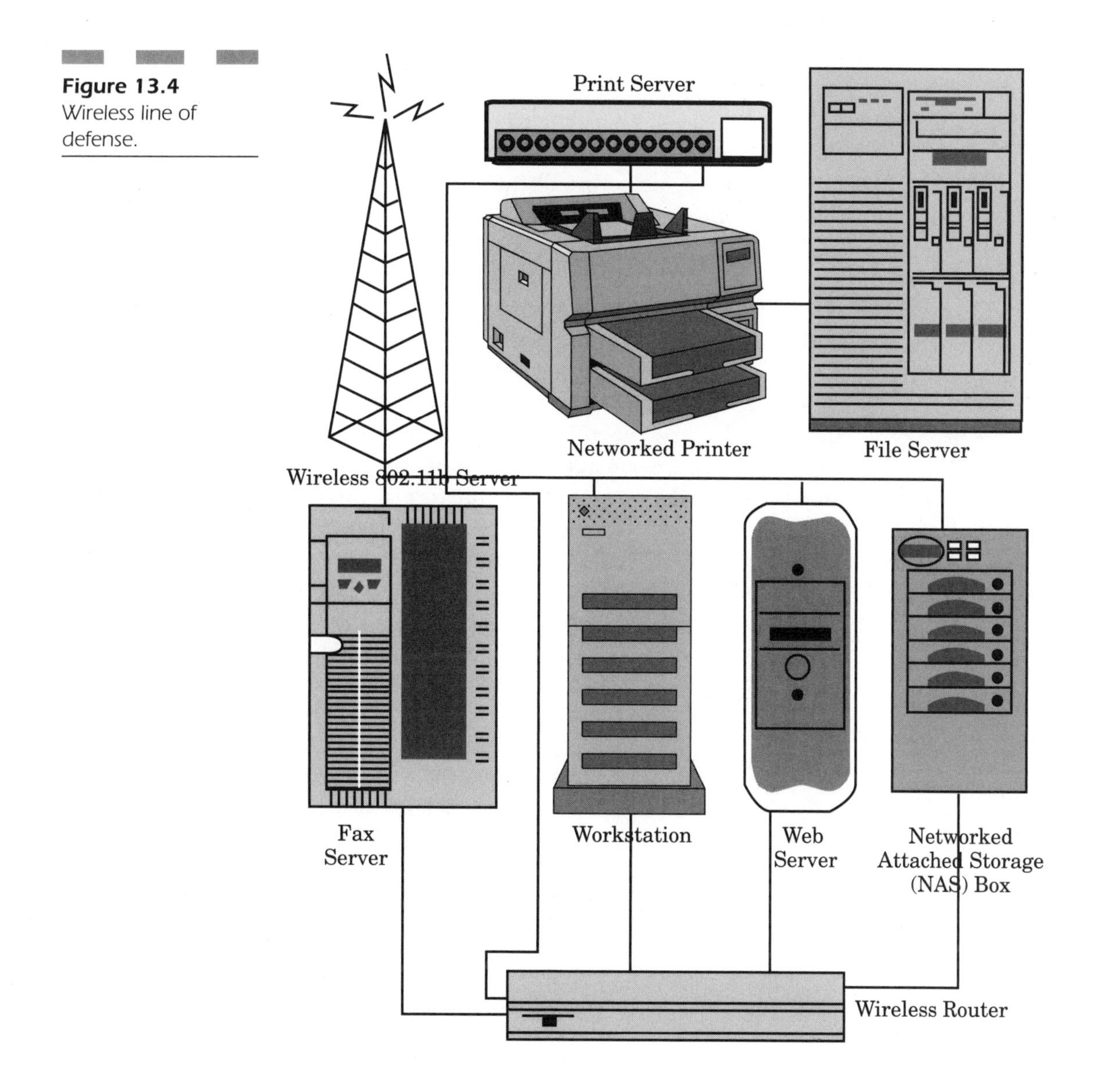

Figure 13.4 Wireless line of defense.

The first step for any networked device is to read the manual and determine how well you can execute the security settings so that very few people know the access codes, logins, and passwords to access the device. In this way, even if someone could see the unit on the network, it would be difficult, if not impossible, to access it.

The most vital concept, of course, is to keep an eye on your internal and external network access points. If you configure your wireless network to accept network connections only from those network cards you trust, then it would not be possible for someone to sit outside your building and set his NIC card to promiscuous mode to try to access your network resources.

In addition, always remember to assign at least some level of encryption to your network traffic so that it becomes that much more difficult for someone trying to break into your wireless network to decode your information.

Taking steps to prevent hackers from eavesdropping on or accessing your network is simple, but requires the time and patience to know these settings exist and then to set them. The rule you should follow is never to put any device on your wired network without knowing exactly what types of inherent security features it offers to restrict access.

Most network printers, for example, can restrict themselves to functioning only in a certain domain and being accessible only to specific users. You should consider restricting access to network devices so that only authorized users can attempt to use these valuable resources. If someone can access a device on your wired network, you can be certain that someone can access that same device on the WLAN too.

One last good measure is to set your network devices to keep a log of all incoming network traffic, most especially traffic received from wireless stations. If all else fails and you don't know how you are being hacked (or the hack is so subtle you don't even realize anything is happening until it is too late), you can use the information in these logs to track down the culprits responsible for disrupting your wireless network. Even if you can't find the people responsible for destroying the integrity of your WLAN, you can at least use this information to plug the security hole in your wireless network so that hackers can no longer exploit open pathways to different devices on your network.

Conclusion: Limiting Your Vulnerabilities

Remember that no matter what device is connected to your network, right out of the box there are few or no security features enabled by default. This is a fact for just about everything on your network. The goal of this chapter has been to point out some of the major possibilities that would cause problems for your wired network from hackers unauthorized to use the wireless network.

First, make sure you understand all the security settings available in any network device in your network. Note that items such as file servers or network printers are attached to your LAN and can be very easily accessed or abused by someone on your WLAN. Understand how to protect these settings; restrict access to those who are directly responsible for the administration of these devices. It is important to note that items on your network that don't seem obvious targets for hackers are vulnerable and can easily facilitate a simple security breach you would not normally have considered.

Two methods of visualization are important when trying to consider how security plays an effective role within your wireless network: internal device security and external network security (Figure 13.5).

Internal device security is applicable to NAS boxes and networked printers. Make certain to set the LAN segment these devices can function on and restrict access with a login and password for each resource. Do not allow functionality to be accessible by "any user," which is the common default on almost every network printer. Note that hackers can deplete your paper and ink and reduce the overall life of these devices by misusing them at all hours of the day and night. Hackers can cause extreme network congestion by sending large graphic files over your network to wait in endless queues to be printed by almost any network printer.

NAS boxes represent easy file access across your network. These common devices are hooked up to your wired LAN, but are extremely easy to access as a public file resource for any wireless user. That means any file, program, or other document on these file servers can be destroyed, corrupted, or stolen by anyone. Note that a wireless user has all the same access rights as a standard internal LAN user. This means your intranet is unsafe and unprotected!

Finally, consider all the external types of access breaches that a misconfigured access point can represent. If you don't plug holes that allow a hacker to use a promiscuous wireless NIC card to attack and breach

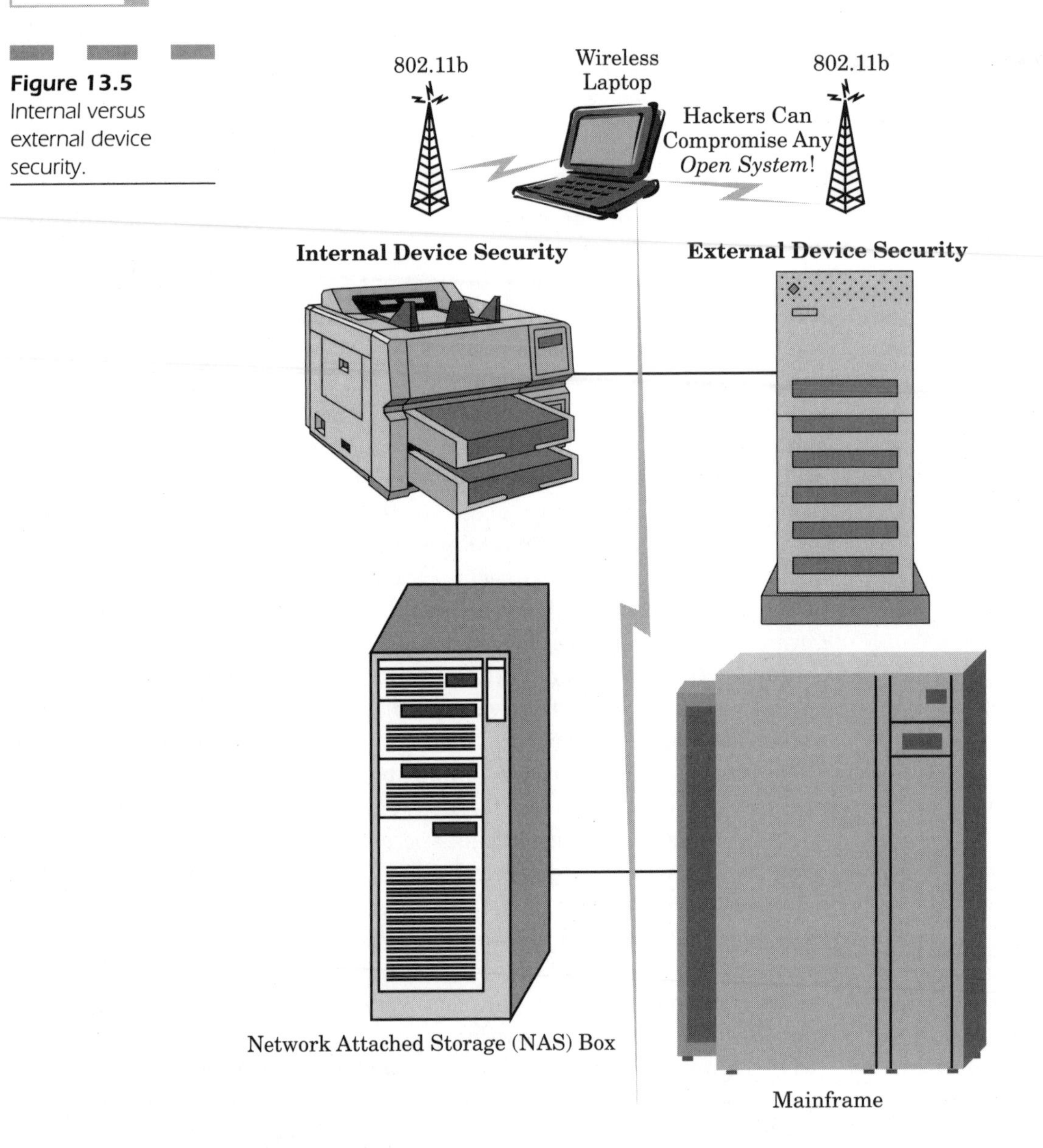

Figure 13.5 Internal versus external device security.

your systems from the parking lot of your corporate facilities, then you are leaving yourself wide open to attack.

Try to think from the hacker's perspective:

- What types of resources are available to just "anyone" who is a wireless user in your network?

- How wide and far reaching is your wireless network?
- How many different users exist on your network?
- Have you registered every wireless NIC card so that you don't allow just "anyone" to access it?

If you consider these questions and more, you can more easily determine how to defend your entire network from a wireless security breach. Once you eliminate as much vulnerability as you can, then you have a far greater level of protection that enables you to survive a hacker attack than someone who didn't read this book and may be unprepared!

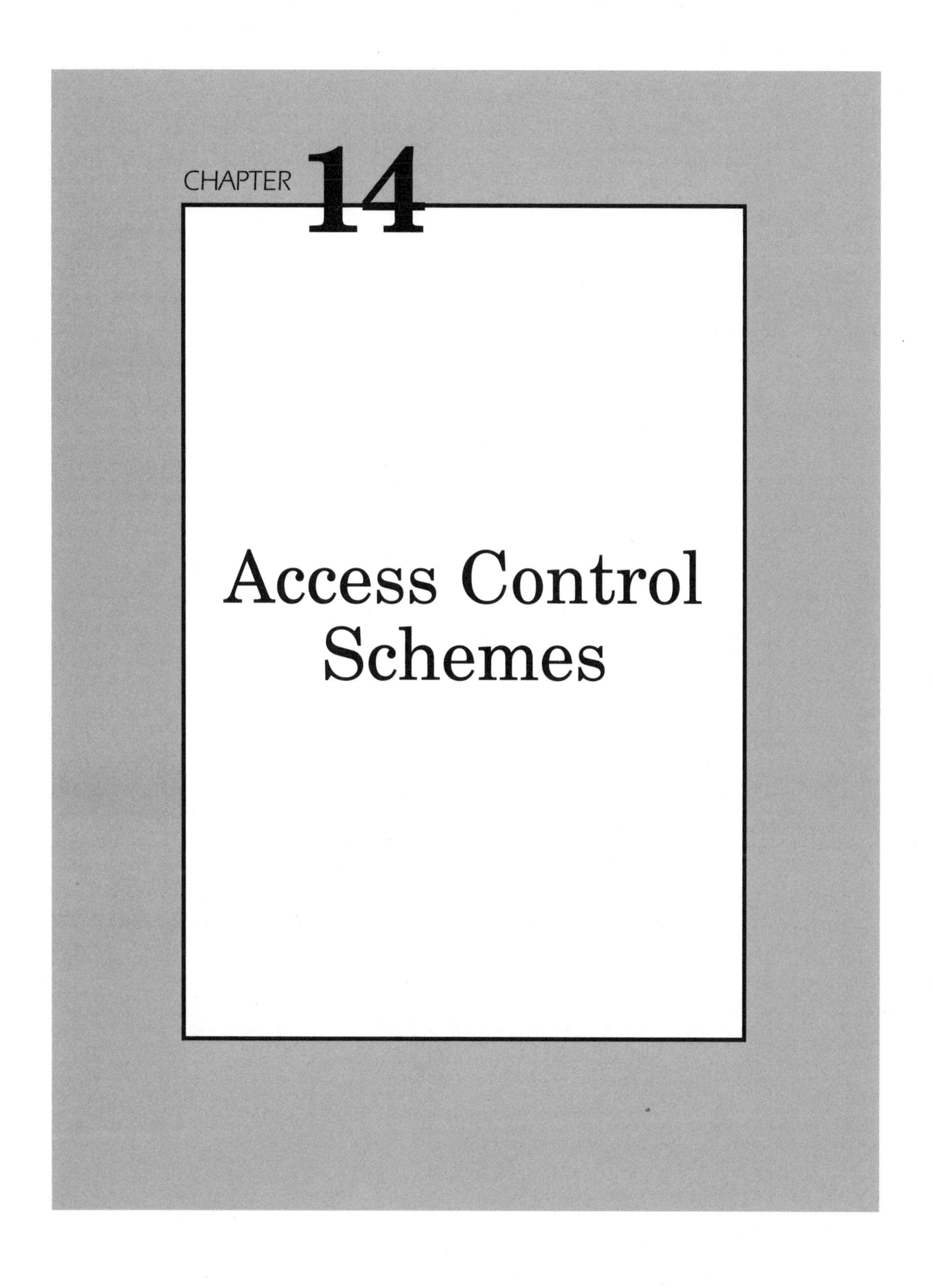
CHAPTER 14
Access Control Schemes

The problem with most access control schemes is the lack of attentiveness by users after they have logged in. This chapter explains common user mistakes in Windows and Macintosh computing environments that open up security holes in their workstations, thus allowing anyone to gain internal network access to mission-critical business systems.

Authentication

The 802.11 standard specifies an encapsulation technique that permits the transmission of EAP packets between both the "supplicant" and "authenticator" within your wireless network.

EAP offers a standard means for supporting extra authentication methods within the PPP protocol. EAP supports several authentication schemes (Figure 14.1), including:

- Smart cards
- Kerberos
- Public key
- One-time passwords

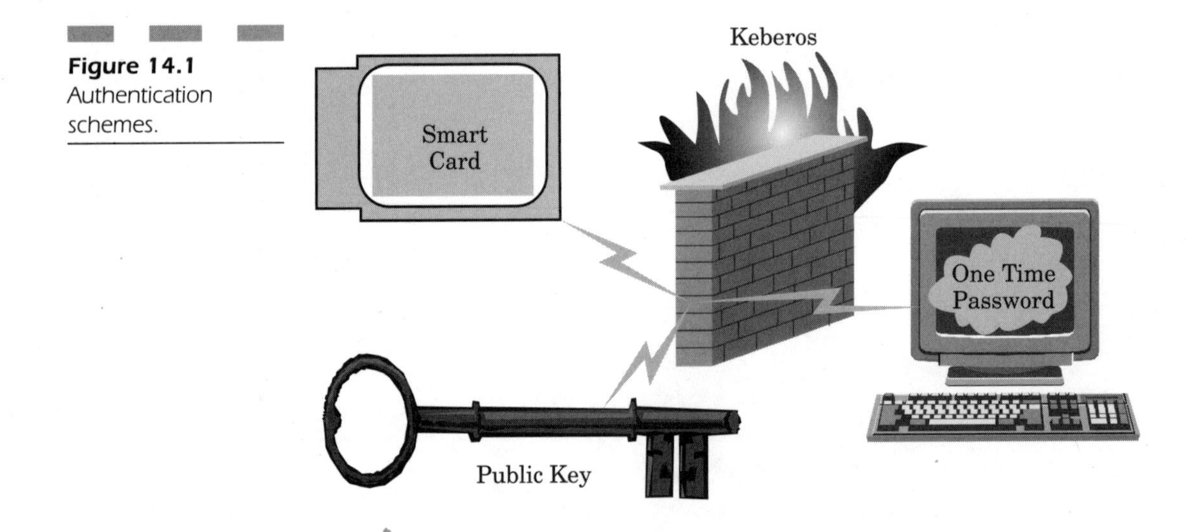

Figure 14.1 Authentication schemes.

Windows XP Access and Authentication Schemes

The platform-specific mechanisms within Windows XP support the following types of methods:

- Username/password
- EAP/MD5 authentication methods
- PKI-founded EAP-TLS

The EAP/MD5 was mainly created to function with EAP, and its use is not usually good for a number of applications. When you use the username/password authentication through challenge/response mechanisms, it is done right over the WLAN. However, this makes it vulnerable to dictionary attacks.

MD5 in and of itself does not offer "mutual authentication"; it only permits the server to validate the client in any given area, but does not have the sufficient client/server instances necessary to decipher keys to create a secure channel of communication.

The EAP/TLS authentication mechanism is PKI based and uses certificates based on or stored in smart cards or the Windows registry file.

EAP/TLS offers the means to have mutual authentication by protecting the integrity of cipher negotiation and key exchange from a sending point to the receiving point. TLS authentication mechanisms allow for mutual authentication that works with client and server so that each is validating the other through special certificates.

Access Control Procedures

There are a number of practical steps you can take to ensure control of who accesses your WLAN. It is important to make certain that you create and maintain a secure 802.11 wireless LAN. When you implement a set of access-control procedures, you acquire a higher level of security for your WLAN.

First, create an organizational security policy that utilizes wireless 802.11 protection features. You must then make certain that all the users on your WLAN understand and can use the security features and functionality that prevent the risks associated with wireless networking.

One way to offer an added level of protection is to perform a "risk assessment" that allows you to comprehend how important your data assets are and how they require protection. You should also make certain that your wireless client network interface card and access point are capable of supporting firmware upgrades and security patches. Both these elements are important; they help you protect yourself against hacker exploits as they become known.

Security assessments must be comprehensive and complete in order to afford sufficient protection. They must be completed at specified intervals and include a sufficient level of validation of all access points connected to your systems. Access controls can be seriously compromised when rogue access points are installed within transmission range of your wireless workstation clients. All this information is crucial in order for you to maintain adequate access control to all the devices on your wired and wireless networks.

In order to maintain proper access mechanisms, you must also carefully examine the external boundaries of your corporate network. Wireless networking devices can have a transmission range that goes beyond buildings. Thus, it is essential that you define the secure areas of your wireless corporate network. If you are careful about the range of your wireless networking devices, you can operate with relative certainty that unauthorized hackers will be unable to access your network by operating a mobile device in the fringe area of wireless transmission without your knowledge.

Physical Security

One of the best ways to make certain that you control access to your computers is to implement mechanisms that control access to the physical areas of your corporate facilities. You should also keep records and restrict access to any sensitive areas within your company where unauthorized access might present a risk to your overall security.

Physical access controls should be deployed in all buildings in your corporate facilities as well as any other secure areas within your organization. Standard physical controls that are essential include having proper identification (Figure 14.2) for each employee:

- Physical ID
- Magnetic or tape badge readers
- Keys that grant only essential building access

Figure 14.2 Physical identification.

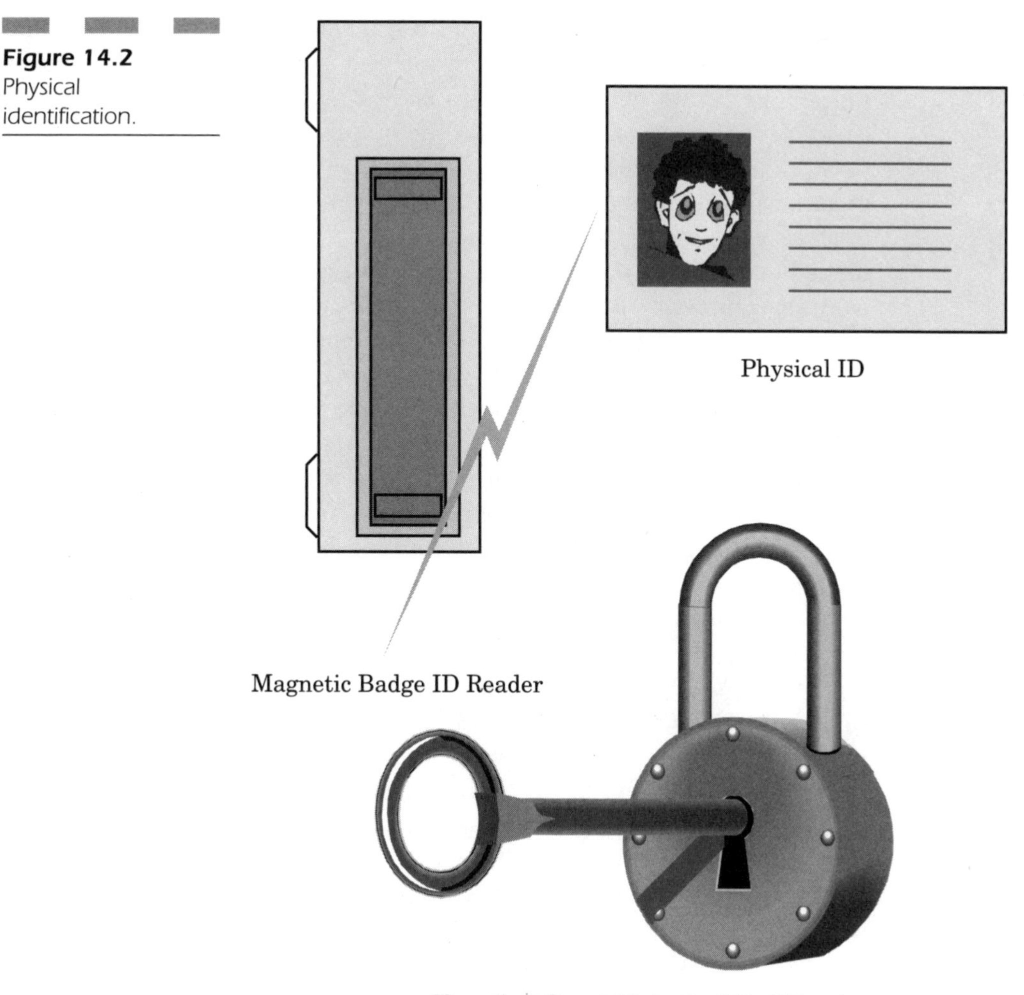

Controlling Access to Access Points

Make certain you have taken a complete inventory of all the access point devices within your organization. Do you know who has physical access to the rooms in which these devices are located? You need to restrict these areas so that nonessential personnel don't have access; this will help prevent their settings being changed to make it possible for a hacker to access your network without anyone even knowing the difference.

It is essential that you maintain these controls by executing a complete site security survey to determine and establish the most effective placement points for any wireless access points so that transmission coverage is limited to areas within the building. If you can log into your network from the parking lot of your facility, so can a hacker just trying to gain access!

Once you have tested your site for these problems, you should also make it a priority to test the transmission range of your network access point. It is imperative that you find out exactly how far your wireless coverage extends, so that you can take the proper measures to shield your signal from undesired areas near your corporate or network facilities.

Another test of the physical limitations of your access points is to determine if any other WLANs are operating near yours. Your site survey should include the ability to test nearby areas for other wireless networks. You can then take appropriate measures to protect your network against the transmissions from someone else's WLAN.

One way to protect your WLAN is to make certain that the access points from your WLAN are at least five or six channels apart from any other nearby WLAN than your site survey detected. One of the problems in having WLANs close to one another is "interference," which may cause undesired reductions in throughput, but also mask a real incoming hacker attack hidden under the interference patterns.

To give yourself better odds at ensuring that only authorized personnel have access to your WLAN, specifically locate your access points on the inside of your buildings with enough shielding on exterior walls and windows to block the signal from straying too far from the building itself.

Physical Access Point Security

Physical security for access points should include not only placing them in secure locations that allow limited personnel contact with these devices, but also programming hourly usage patterns so that access points are "turned off" during non-business hours. Hackers realize that most companies leave these devices on all the time. This is just an invitation to hackers to use these WLANs to break into your network during off-peak hours, when it is far less likely they will be detected by anyone on your staff.

Another method of detecting malicious activity to your access point devices is to monitor them to see that their "reset" feature is only being used when an administrator needs to reinitialize the device. Only authorized administrative personnel should have the power to change these sensitive settings; otherwise if a group of employees has access to these devices, it is simple for them to be misused or reprogrammed to allow unauthorized users access to their information.

One of the problems with resetting access point devices is that every time you reset the device, you can do so with an entirely new set of security settings that can compromise the integrity of your WLAN and allow unrestricted access into your network.

Secure Access Point Management Issues

When you buy an access point, never make the mistake of thinking that the device is already secure out of the box. One of the most common mistakes people make is that they never think to change the *default* SSID in the access point. Most hackers attempt to gain unauthorized access into your device just by knowing what your SSID is set to.

Another way that hackers gain access into your system is to set an SSID field to a "null" value. That way, the machine will automatically look for any access point that broadcasts its SSID and immediately log onto the network, whether or not the hackers know what your SSID happens to be. In order to protect yourself, it is imperative that you disable the feature that allows you to "broadcast your SSID," so that the client must know your preprogrammed value before trying to access your system. At the very least, this measure protects you from prying eyes or hackers in the fringe reception areas attempting to log into your network.

SSIDs are very much like passwords (Figure 14.3). Hackers try to guess the value of this field by looking at simple items of information that include:

- Company name
- Division name
- Department
- Street location
- Your name
- Your product name

Figure 14.3 Information appealing to hackers.

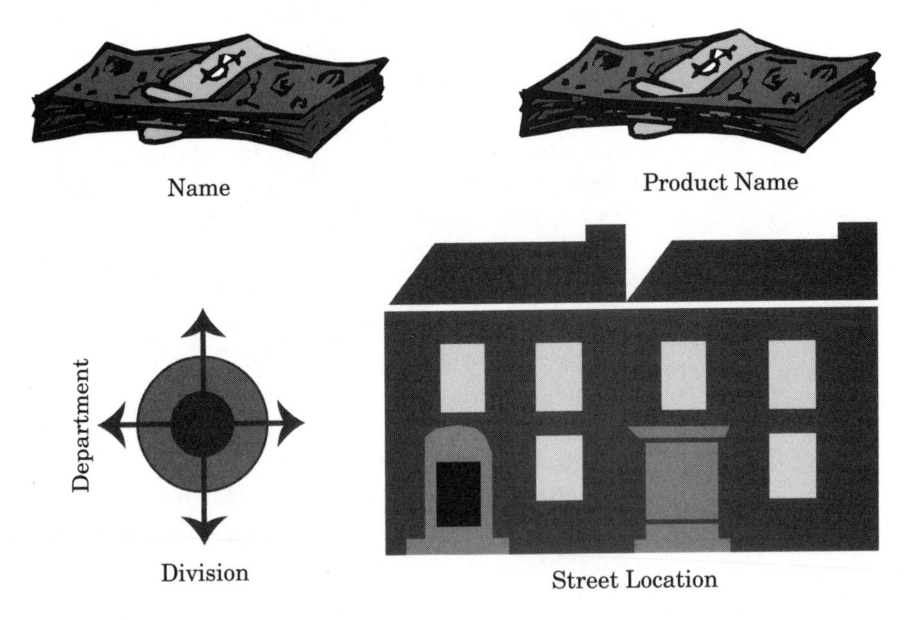

People never think to set the SSID using the same secure rules they use when establishing a secure password. When a hacker does try to break into your system, the first thing he or she will do is attempt to guess your SSID by using familiar information as described above. This is why it is imperative that you restrict the SSID field to a value that is difficult to guess from any information that tells people about you, your company, or what you do.

Another way of protecting your wireless access is to disable the broadcast beacon of your access point. Don't allow your equipment to advertise that you have a WLAN that can be accessed. This information tells hackers enough information about your network to give them the edge they need to breach your security and gain access to your WLAN.

In short, *change all the default* settings for your access point. Any default value is a potential way for a hacker to access your internal network, change settings, or cause havoc that would prevent authorized or legitimate users from using the valuable network resources they need to do business. Note that there are a wealth of hacker sites, freely accessible on the Web, that index every single wireless router password and setting. While these values are used primarily for people interested in security, they are easily misused by hackers who can access your network resources simply by knowing something as simple as the vendor and model number of your wireless networking hardware.

When you set the options for your access point devices, you must make it a point to disable any service that does not pertain to your business operations. Any nonessential management protocol is a potential risk and adds to the insecurity of your wireless network. Any service left on that is not being used is another invitation for a hacker to gain access to your network through wireless exploits.

In contrast, it is imperative that you enable all the security features and functionality on your WLAN that dissuade a potential hacker from attempting to access your systems (Figure 14.4). These functions include:

- WEP
- Privacy features
- Cryptography
- Access control
- Authentication mechanisms
- VPNs

When you do enable encryption keys, you should use at least a 128-bit key or larger if technology permits. If you use any "default" shared keys, you should consider trying to replace them with "unique" keys that have greater levels of security and are much harder to guess if a hacker tries to exploit that vulnerability in your system.

In addition, default shared keys should be replaced by unique keys on a regular basis. These types of keys often fall into the wrong hands and can represent a security vulnerability in your WLAN if they are not changed often enough. It is imperative that you establish a policy to

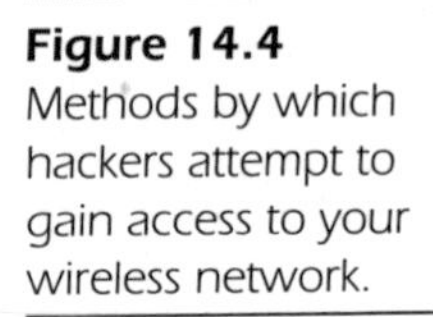

Figure 14.4 Methods by which hackers attempt to gain access to your wireless network.

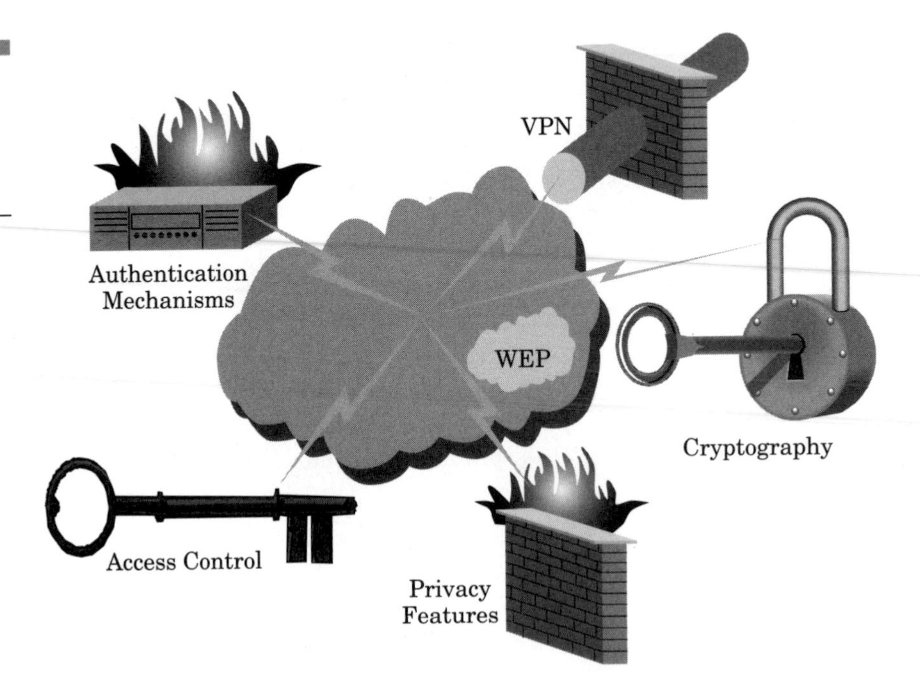

make certain your keys are replaced often, and that you have a set of rules to define how unique these keys will be to make them that much harder to guess.

Establishing a wireless firewall After you have carefully examined your WLAN and have taken every precaution possible to control the access mechanisms to any potentially vulnerable parts of your network, you can install a carefully configured firewall that is deployed between the wired Ethernetwork within your organization and your wireless network near your access point. This type of protection is an essential element that most companies neglect to install.

The WLAN is the most vulnerable part of your network because anyone can access resources without being physically within your corporate office buildings. If you install a firewall, then you can effectively block off access from any incoming wireless client into the protected resources of your wired network. This level of protection is essential and allows you to make certain that if your WLAN access controls do fail, you have the firewall preventing any malicious or unauthorized users from gaining access to resources that should most definitely stay *off limits* to all unauthorized users.

Preventive Measures

One common ways that hackers can gain access to your wireless network is to send a "Trojan horse," or a program that makes it possible to circumvent your access control schemes by infecting a file with a virus. Your best protection is to install and constantly update the virus definitions at every wireless workstation.

Without your ever realizing it, there are a number of ports and open services on most workstation clients. Many clients have Web, FTP, and mail servers enabled. All these "available services" are invitations for a hacker to gain entry into one client and work his way through your entire system. The best means of protection is to install a personal firewall on each one of your wireless clients. This is the most effective way to make certain that anyone who does try to access any services on your wireless workstation won't be able to get past the firewall blocking these ports.

MAC the Knife

A good preventive measure that stops a hacker from gaining access to your WLAN is to keep a MAC access control list. This list identifies the unique code of each networking card. The server or access point can be easily programmed to either grant or deny access to an incoming network connection based on whether or not that code is in the acceptable list. Most hackers gain access because this feature is not enabled by default. In an open networking environment, anyone can gain access to your system at any time. The MAC access control list (ACL) screens out any unauthorized user attempting to access your WLAN. This simple measure is a very effective means of protection.

VPN

Because the WLAN inherently offers poor encryption techniques to protect your transmission, you may find that deploying an IPsec type of virtual private network (VPN) will further enhance your ability to protect your transmission, through encryption. If a hacker does find a way to decipher the WEP encryption of your WLAN, it won't do him or her any good, because the underlying transmission has its own encryption algorithm. This is really the best method of making certain your transmissions are not intercepted by any malicious user.

When you implement encryption, it is imperative that you use the highest strength possible, due to the confidential nature of network data. Since computers have become faster and more powerful, it is much easier for these devices to decipher your encryption algorithm. The less powerful your encryption scheme, the easier and faster it is for a computer to decipher your wireless network transmission.

Patchwork Sometimes it almost seems to be a full time job just trying to stay up to date with all the software patches and upgrades available. Fortunately, these software patches serve to resolve the security problems found in your software and operating systems, which seem to arise on almost a daily basis. It is imperative that you both test these patches (to make certain that they work) and deploy them on all your wireless workstations at periodic intervals to protect your network against hacking attempts.

Passwords Password rules are commonplace in most IT environments today. These rules should also extend to the administrative passwords of your access points. Hackers are always looking for the opportunity to use a "dictionary attack" to find an easy password to the routing device in your organization and alter its settings to permit unrestricted access into your corporate intranet. You should also remember how important it is to change passwords on a regular basis. Passwords become stale very quickly, so you must stay ahead of the game by making certain you change them regularly so that someone doesn't find out how to gain access into your systems.

Enhanced access-control schemes Make it a point to deploy special types of user authentication modules to control access into your system. These types of elements, which should be standard on your WLAN (Figure 14.5), include:

- Smart cards
- Two-component authentication
- PKI
- Biometrics

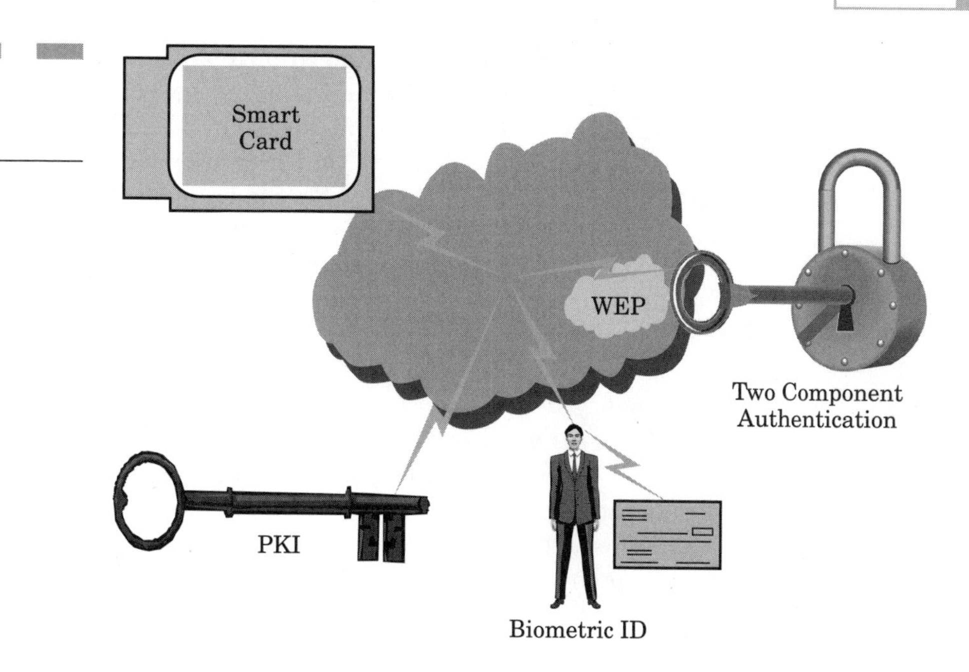

Figure 14.5 WLAN core components.

IP Addressing Issues

IP addresses are a common way for a hacker to attack your network. It is imperative that you make certain that the ad hoc mode has been disabled in your WLAN, since it constitutes a risk that an unauthorized user can log into your network.

If all your workstations are behind a router, you should assign a static IP for each wireless workstation so that you know exactly what IP address is assigned to each person within your organization. If a hacking attempt does occur, a quick examination of your log will tell you which workstation has been compromised.

Security concerns that open up a hole in your access control mechanism involve having DHCP addressing. Even if you have taken the above statement to heart and have assigned a static IP address to each worker within your corporate LAN, it is still possible to use DHCP. Most administrators enable DHCP for convenience and program the wireless router to assign the same IP address to each workstation based on its unique MAC ID. Unfortunately, it is all too easy for DHCP to assign a dynamic IP to a user who is not on the predefined list. This vulnerability is commonly overlooked because this feature is “on” by default in most access point devices.

Managing administrative functionality Hackers look for ways to access the management or administration pages of access points so they can enable the DHCP server and block out any ACL that might be preventing them from logging into your network. One of the ways in which you can protect yourself is to enable the user authentication features of your access point so that you can make it exceedingly difficult for hackers to log onto these devices or change their settings.

Another means of making certain that hackers can't use wireless channels to alter the management settings for your access point devices is to make certain that all traffic designed to control the management functionality of your access point can only be accessed through your wired Ethernetwork. In this way, a hacker would have to gain access to the actual wiring inside your company in order to do any damage. Since most hackers only look for resources they can access through your wireless network, controlling the access to local workstations provides a little more protection.

Another way to control access to your wireless router is to use a serial port connection interface, as opposed to a network connection. The idea is to reduce the possibility that someone will be able to reconfigure the device to access sensitive or confidential information within your organization.

Alternative authentication You may choose to implement alternative forms of authentication for your WLAN including Kerberos or RADIUS.

If you rely on the protection capabilities of your wireless equipment, then you will have enormous difficulty in finding answers. Don't depend on WEP; you should consider your WLAN as an open network (even if it is not), so that you can implement better encryption policies and other types of authentication safeguards that will allow you to control inbound access to your wireless network.

Finally, look into deploying intrusion detection systems on your wireless network to ensure that you have the power and the capabilities to detect any malicious hacking exploits. The idea is that by knowing who attempts unauthorized access attempts and hacking activity on your WLAN, you will have sufficient information to make certain your network resources can catch these attempts, alert you to these problems, and allow you to sever a connection before it compromises the integrity of your wired and wireless networks.

Conclusion: Ensuring "Secure" Access Control

In order to maintain secure access control of your wireless network, you need to think proactively about the deployment of your wireless products. Think about the security functionality and the specific features that are important to enable to provide you with the protection you require, including authorization functionality and cryptographic protection.

Your goal is to comprehend fully the specific impact of deploying a security solution function before you actually put it into practice. There are a great many permutations that could result from any security product. Some of these products, if misconfigured, could constitute a security problem instead of providing you with the protection you require against hackers.

When you deal with your organization, it is important to have one central person responsible for identifying your security features and the functionality of your wireless security products. This person can deal with either potential security threats or the vulnerabilities of your technology.

As security is a constantly evolving science, wireless standards will evolve to incorporate new features, enhanced functionality, and protection against constantly changing hacking exploits. If you are careful and pay attention to the ever changing face of the security industry, you can effectively act to prevent any breach in your access control schemes and maintain the integrity of your wireless network now and well into the future.

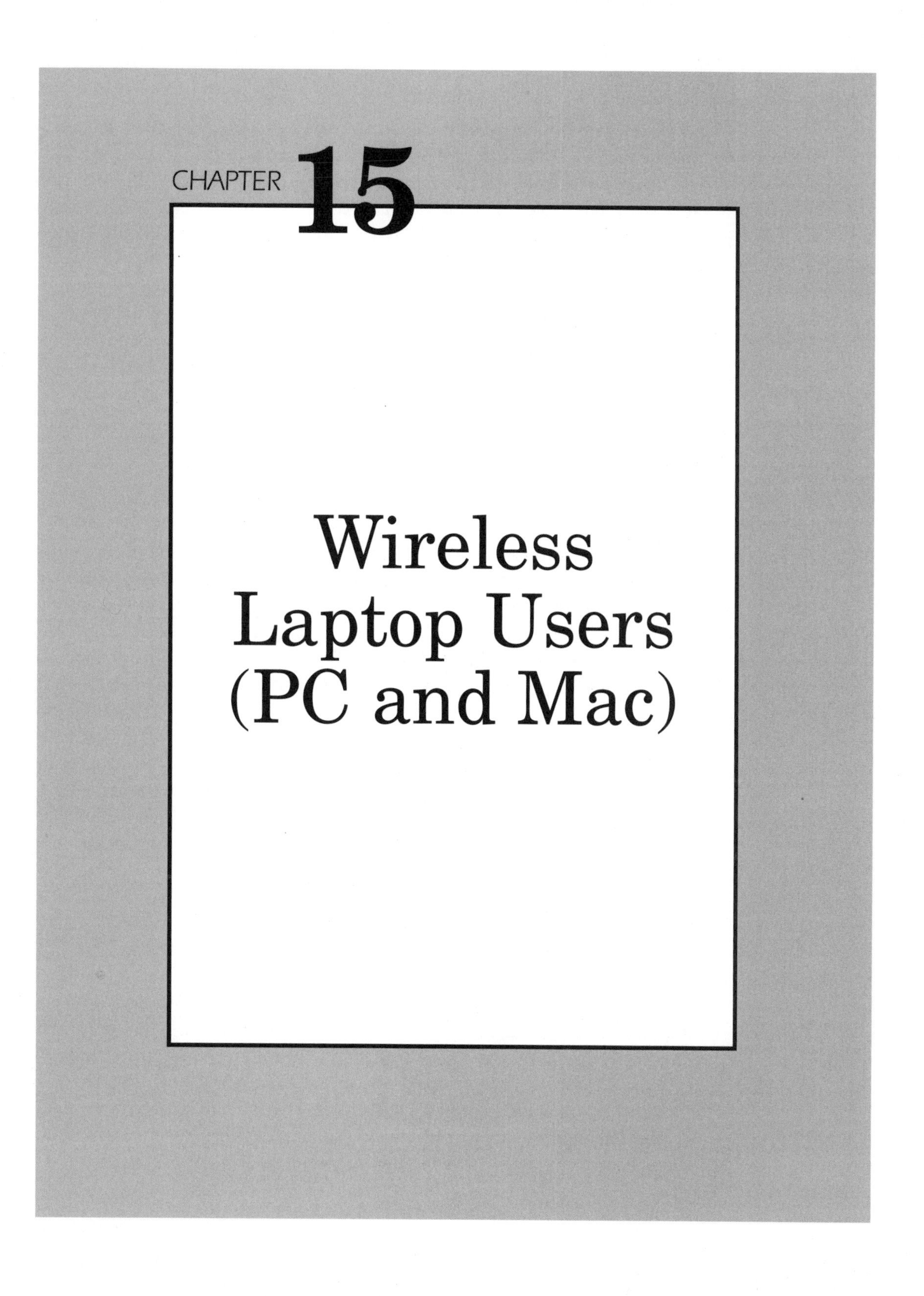

CHAPTER 15

Wireless Laptop Users (PC and Mac)

Wireless laptop users can leave a connected computer unattended while on the road or within their offices. Laptops are often confiscated and left as open portals at airports and other public places. This chapter describes how 802.11b is built into many laptops and is already configured to access network shares. They are vulnerable, and most users don't even password-protect this precious resource.

Laptop computers represent a significant and prominent security threat. This mobile device has the greatest vulnerability of all your corporate business applications. This chapter will examine the threats these critical devices pose, how to mitigate risk, secure WLAN applications, and reduce your vulnerability.

Laptop Physical Security

The laptop computer is the best and worst device you could possibly own, at least from a security perspective. It is the best device in terms of sheer power and portability. However, it is the worst device because it can be easily lost and represents a serious gap in your security.

Physical security of your laptop primarily deals with unauthorized users who acquire your computing device and use it to eavesdrop on or access wireless network communications. The problem with losing a laptop is that these types of intrusions are difficult to trace. They could be coming from inside the network or outside, if your access point has sufficient transmission power and extends beyond the building perimeter. A hacker could use a stolen device to listen to your internal network traffic from the parking lot outside your building without your even knowing about it.

Protection

You can take special steps to protect yourself against the possibility of any wireless laptop being used against you. You need to implement encryption to limit the eavesdropping attempts against your network communications. Some companies use a special VPN link that changes its value every minute or two. If someone steals your laptop, it wouldn't do the hacker any good to try to access any network resources without the proper decryption device to log into your network.

Eavesdropping attempts on a wireless network can take place from areas that are in range of the access point, but the risk rises significantly when there aren't any enabled encryption parameters set for the access point itself. Hackers are looking for devices that don't even have basic protection settings enabled, because they are the easiest to break into.

MAC access control lists Hackers are most interested in getting one of your employees' laptops because it is easy to steal and is already enabled on your internal access control list (ACL). One of the problems with just trying to access your WLAN is that most administrators are wise to the fact that they can set their wireless access point or router to screen out any wireless network interface card that does not match their ACL of preprogrammed unique identification values. If a laptop with an unknown MAC address tries to log into the network, access is denied. However, if a hacker steals a corporate laptop, he can take the wireless network interface card out of the computer and use it on another machine in an attempt to breach your security and access your protected network resources.

There are certain limitations to using MAC addresses. Not only can these cards be stolen, but the address itself is transmitted in clear text without encryption over the WLAN. This makes it very easy to spoof these types of addresses and find still another way to breach your network. By understanding the limitations of these security protocols, you are better able to protect yourself against hacker exploits.

Hardware Solutions

Another excellent way of adding protection to your WLAN is to set up your laptop so that it must identify its legitimate user correctly before the machine will turn on or allow any access to protected network resources. These hardware solutions rely on physical attributes as a means for authentication; these cannot be easily duplicated by any unauthorized users. Using these solutions allows you to reduce your risks of using a wireless network and gives you extra power to ensure you know who has access to your networking equipment.

The primary types of hardware countermeasures are shown in Figure 15.1 and include:

- Virtual private networks (VPNs)
- Public key infrastructure (PKI)
- Biometrics

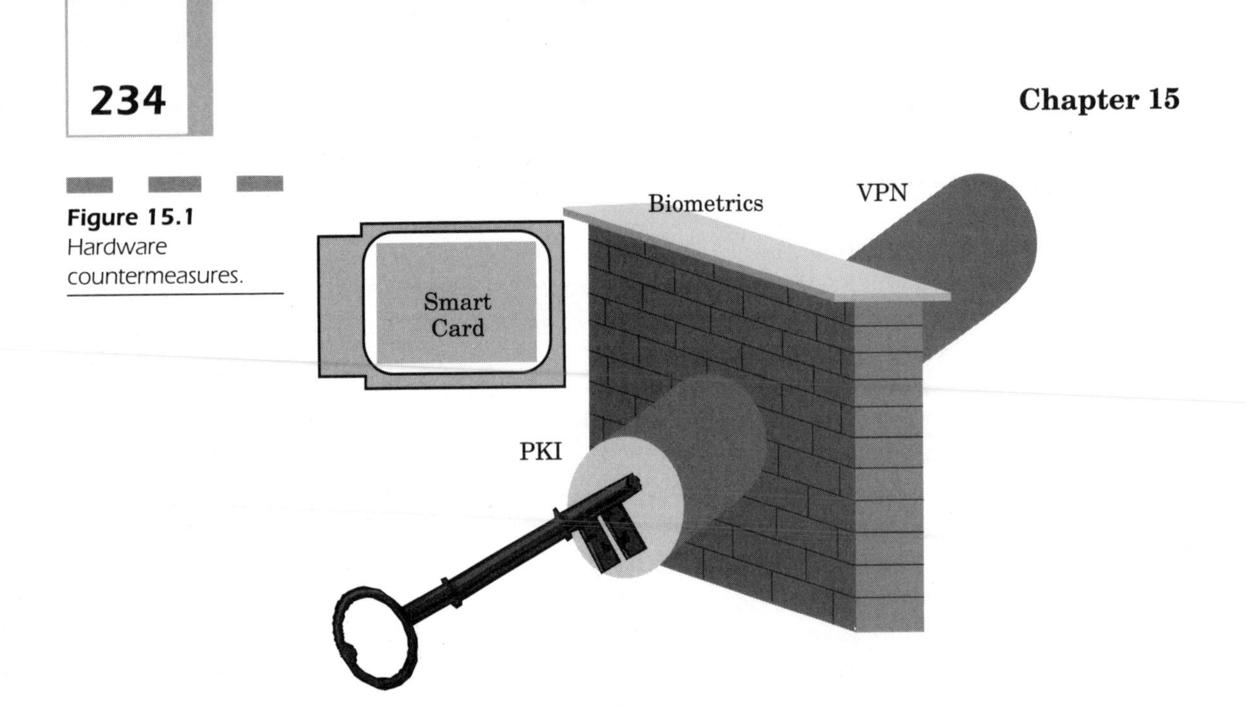

Figure 15.1 Hardware countermeasures.

Smart cards Smart cards are often an effective means of adding enhanced protection to your wireless laptop, though they add another layer of complexity at the same time. When you use a smart card in combination with authentication techniques that rely on your username or password, you have a greater chance of making certain your computer access remains secure. Smart cards can also work together with biometric devices that depend on the physical attributes of the user to access your wireless network.

In a typical WLAN deployment, smart cards offer the enhanced functionality of tighter authentication. They are practical in networking environments that require authentication techniques beyond just a username and password. *User certificates*, for example, are actually stored on the cards and often require that the user know only a special PIN number which can either remain static or dynamically change according to a special algorithm set by a special device.

Smart cards follow the user and are not tied to a specific mobile computing device. They are a good authentication solution that is tamper-resistant for the most part. When you integrate them into your WLAN, you greatly enhance system security.

As with any security solution, it is important for users to understand that smart card security solutions are not a cure-all for the limitations and restrictions of 802.11 security. An effective security solution relies on a number of access safeguards, and the more you use with laptop computing devices, the better off you are.

Virtual private networks for mobile laptop users VPNs provide an optimal solution to secure data transmission over public network infrastructures. They are also useful for security in open wireless networks. WLANs are insecure by their very nature, but when you implement a VPN you add a layer of security that protects your wireless transmissions.

The mechanics behind VPNs utilize cryptographic methods to protect your IP information as it flows from one network location to another. A VPN actually creates a "virtual" tunnel that encapsulates one protocol packet within another. This information is encrypted and isolated from all other network traffic (Figure 15.2).

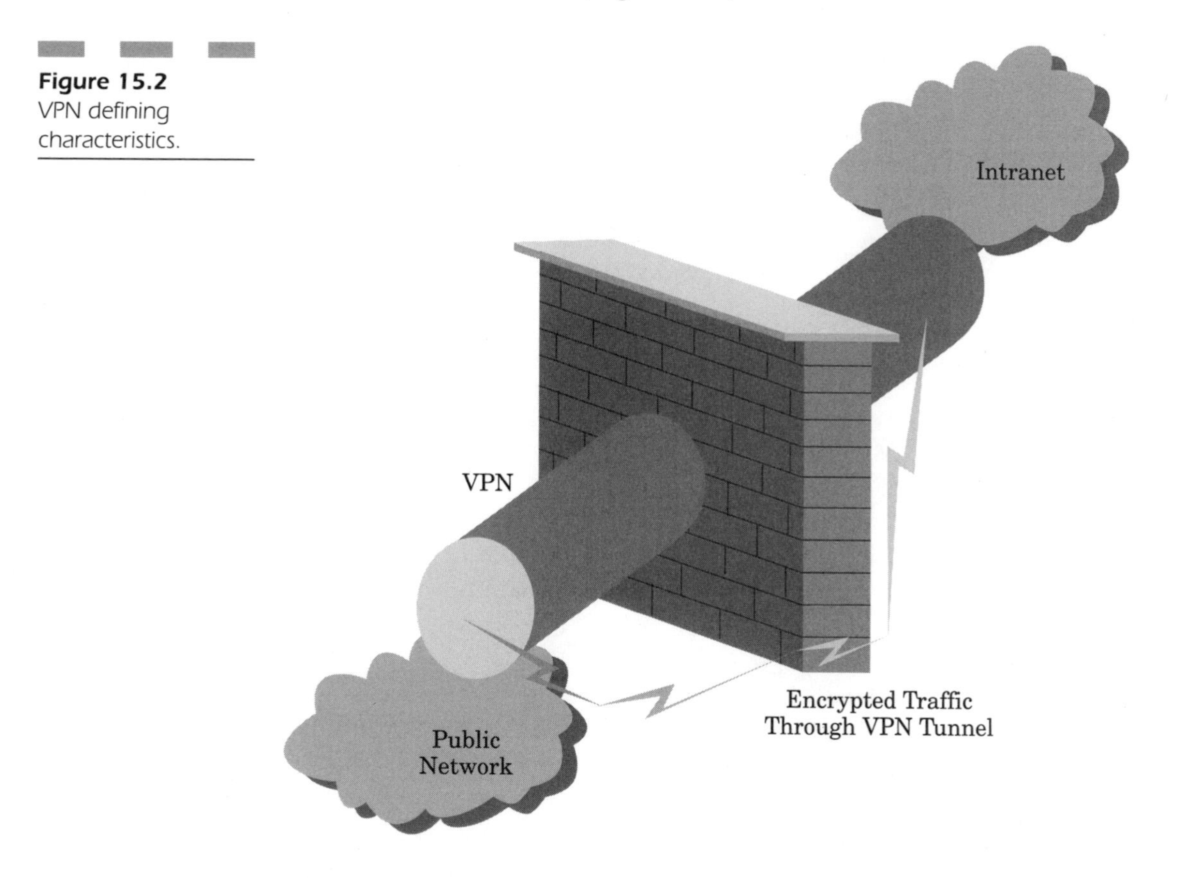

Figure 15.2 VPN defining characteristics.

IPsec is the protocol most widely used by most VPN deployments. The mechanics behind the VPN ensure the following:

- Confidentiality
- Replay protection

- Traffic analysis protection
- Connectionless integrity
- Data origin authentication
- IPsec
- Encapsulating security protocol (ESP)
- An authentication header (AH)
- Internet key exchange (IKE)

Confidentiality Confidentiality makes certain that other people are not able to read information in your private messages.

Replay protection Replay protection gives you the confidence that the same message is not delivered several times. It also ensures that messages are not processed out of order when they are finally delivered to their intended destination.

Traffic analysis protection Traffic analysis gives you the protection you need to make certain that someone trying to use wireless channels to eavesdrop on your transmission is unable to read the contents of your messages.

Connectionless integrity Connectionless integrity ensures that whenever you receive a message it has not been modified from its original format.

Data origin authentication Data origin authentication ensures that the message you receive was actually sent by its originator, as opposed to someone spoofing the information from the person who actually sent it.

IPsec IPsec makes it possible to perform routing tasks on messages through an encrypted "tunnel" using two unique IPsec headers that appear just after each IP header for each message.

Encapsulating security protocol ESP is a header that offers the privacy you need to protect you against any possible malicious attempts to tamper with your wireless data transmission.

Authentication header AH provides you with protection against tampering that would compromise your privacy.

Internet key exchange IKE provides you with the means to permit secret keys as well as other confidential parameters that require protection to be exchanged just prior to the time communication is exchanged. This process works without any user intervention.

Public Key Infrastructure

PKI is an effective way for a laptop user to ensure the integrity of the wireless transmission as well as know who sent the message. PKI yields the services necessary for the creation and deployment of public key certificates. It gives applications the ability to benefit from secure encryption, as well as the authentication of wireless network transactions, while maintaining two important aspects of the connection: data integrity and nonrepudiation.

The benefit of using public key certificates is that WLANs can easily integrate PKI to ensure authentication, with the goal of keeping secure network transactions. In fact, wireless PKI, handsets, and smart cards all integrate effectively with wireless networks.

PKI is an essential element in maintaining higher levels of security, as it offers stronger authentication user certificates. Authenticated users can utilize those certificates with application-level security for both signing and encrypting messages through "encryption certificates."

When these types of certificates are "integrated" directly into the smart card, you have a greater level of security and privacy protection.

However, if your security needs aren't as mission critical as a government project, then PKI may not be the proper solution to secure your wireless network. The drawback of this powerful mechanism is that it is very complex to implement, and there is a higher cost for both deployment and administration. In addition, there are a number of added safeguards that must be taken into consideration before it actually becomes practical for users to adopt a PKI solution for generic wireless networking needs.

Portable Biometrics

Since the laptop is easy to lose and is often stolen, one way to make certain that unauthorized users cannot access your private wireless network is to ensure that they cannot access your computer in the first place. One of the most effective methods of securing your laptop computer is biometric technology.

Biometric devices, as shown in Figure 15.3, include the following:

- Fingerprint scanners
- Palmprint scanners
- Optical scanners (retina and iris)
- Voice recognition

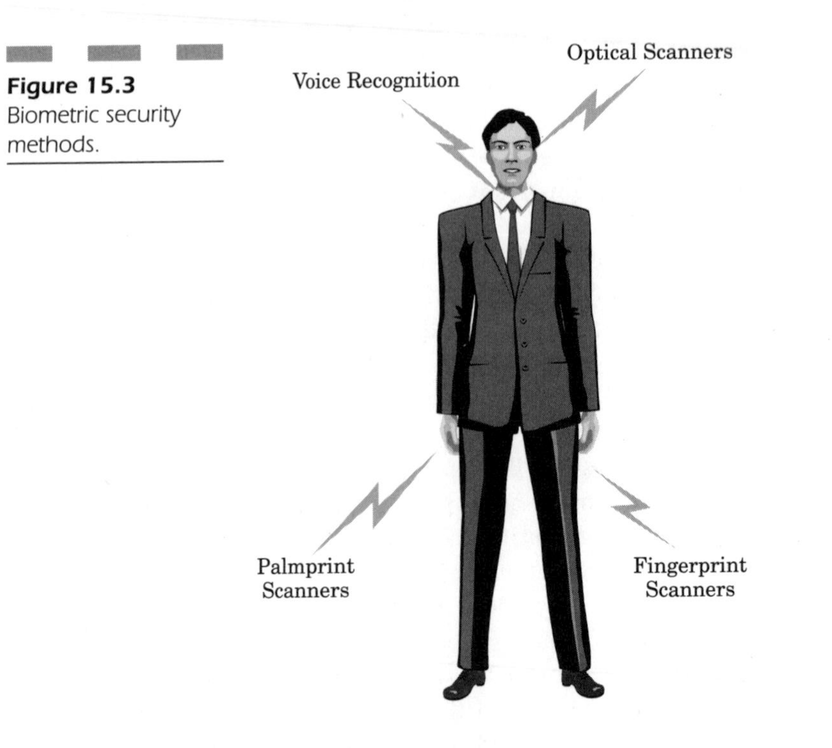

Figure 15.3 Biometric security methods.

If you really want to make certain that nobody can access your PC or Macintosh laptop computer, you could use biometric forms of access control along with other security solutions such as:

- Wireless smart cards
- Wireless authentication mechanisms
- Personal forms of identification that replace the traditional username/password
- Biometric plus VPN solutions (as described earlier in this chapter)

All of these methods combine to yield enhanced levels of authentication and greater levels of data confidentiality.

Reducing WEP Vulnerabilities

There are a number of vendors bringing to market hardware add-on solutions designed to help you overcome the vulnerabilities so prominent with WEP failures within the 802.11b WLAN security space. These products deal with these vulnerabilities in an effort to provide a combined (and more secure) solution in one central product.

BlueSocket One vendor of an effective hardware solution is BlueSocket, which has produced a wireless gateway that establishes a firewall between the access point and the corporate intranet. This device requires that authentication take place through either its internal database or a central corporate server.

This device also supports "central" authentication (Figure 15.4) through its support of:

- Lightweight directory access protocol (LDAP)
- RADIUS
- Windows NT 4 domain
- Windows 2000 active directory
- Extensible authentication protocol (EAP)
- Token-based authentication

You can also use specific roles to assign various encryption levels for each user depending on his specific need for security. When you assign roles, you can also support a level of maximum bandwidth for each user category. In addition, you can support "strong encryption" to deal with the weaknesses of WEP.

Vernier Network Vernier Network has created a system that is actually two hardware devices able to protect your WLAN with the following functionality:

- Authentication
- Control
- Redirection
- Logging of network traffic (respective to each WLAN)

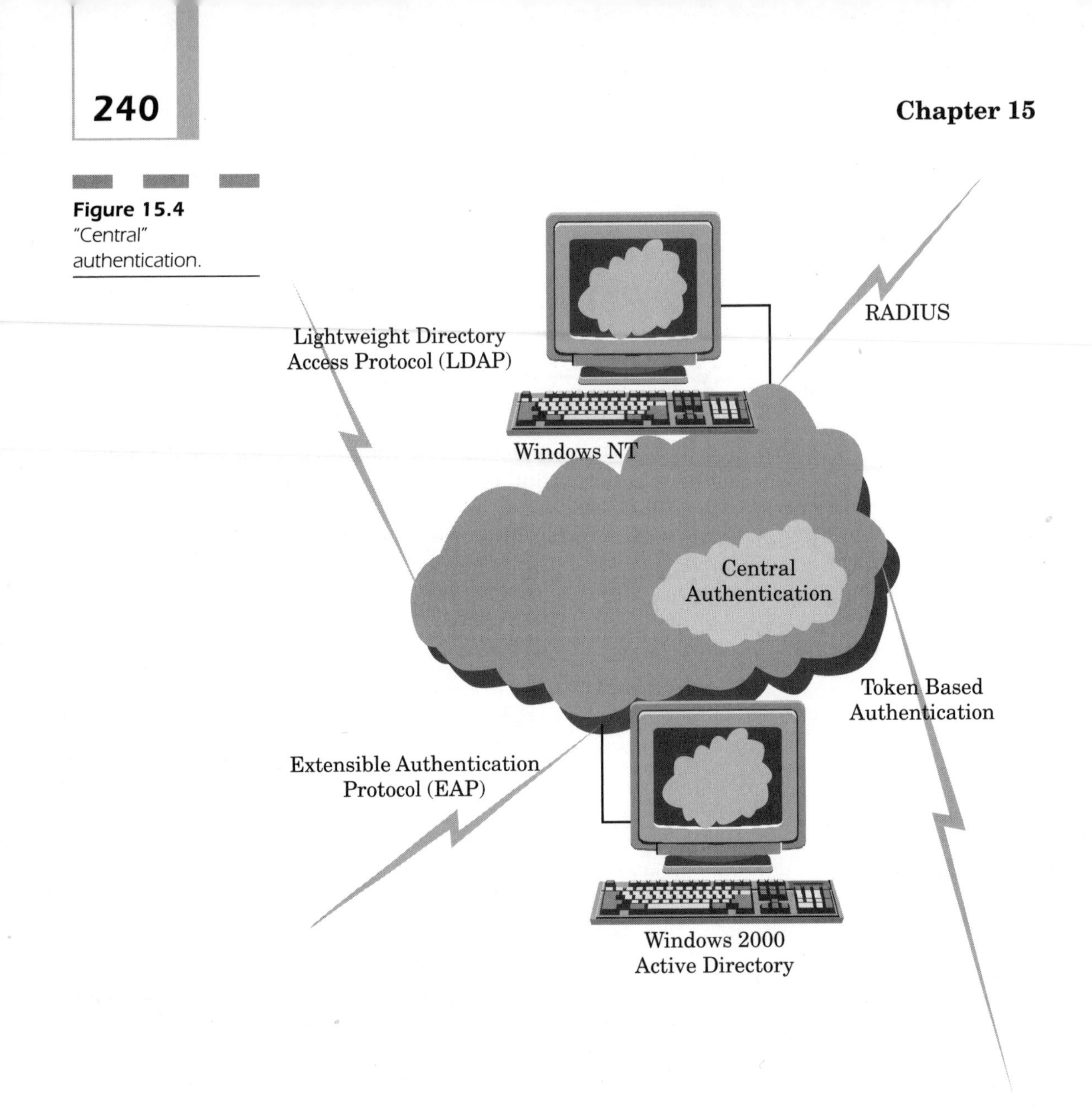

Figure 15.4 "Central" authentication.

These two devices are a control server and an access manager.

The solution is designed to add safeguards to your WLAN that protect against laptop or mobile devices attempting to connect to your WLAN. Realizing that the inherent security measures of the typical WLAN are not sufficient to protect your network resources, this solution attempts to provide you with that mission-critical protection.

The control server is able to manage authentication at one central location for all wireless users and to account for roaming and policy enforcement.

The access manager sits on the perimeter of your network and is able to connect your access point devices. The idea is that it can enforce the user rights for authenticated users. You can also enable roaming, as well as security features and functionality including:

- IPsec
- Point-to-point tunneling protocol (PPTP)
- Layer 2 tunneling protocol (L2TP)

Securing the WLAN

The products in the previous section help to secure your WLAN environment regardless of the weak protection afforded by 802.11b standard.

No matter what solution you decide to implement, it is imperative that you fully examine all your options so that you can make the most effective decision possible when it comes to implementing the most appropriate security features (Figure 15.5) to achieve the following objectives:

- Reduce your risk
- Apply countermeasures to protect your WLAN
- Add enough security to allow authorized users in, but keep hackers out

Platform Bias

PC laptops usually run some version of Windows or some flavor of Linux or UNIX to access your WLAN. Hardware solutions like the one described above are usually your best route, rather than relying explicitly on software to create a protective VPN link. Macintosh and PC computers do communicate on the same 802.11b frequency, but Windows and Mac employ different platforms, which means different encryption software. It is far too easy to fall into a trap in which one platform does not have the most up-to-date version of encryption software, or worse yet, having the Mac not equipped with the proper software to access the VPN that your PC can!

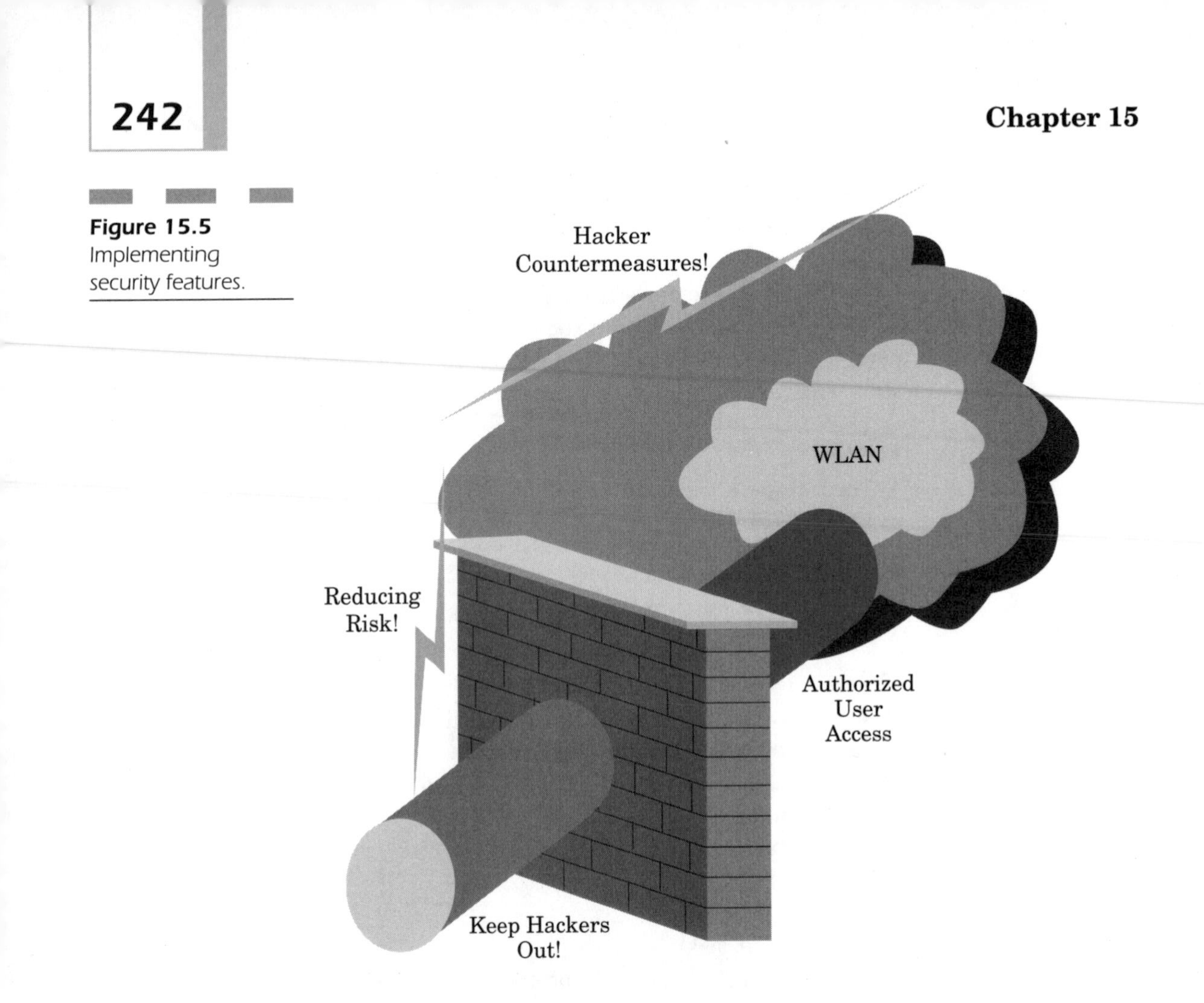

Figure 15.5 Implementing security features.

The point of vulnerability is at your access point, but a hacker can easily find some way to steal your laptop and determine how to break into the WLAN. If you protect your access point by deploying an effective WLAN firewall, *and* immediately deactivate the access privileges from a stolen laptop computer by removing its unique MAC address from the access control list of your network, you can at least have the minimum level of protection required to ensure hackers won't compromise your security and breach your safeguards.

Wireless Laptop Network Support

Windows XP, Lindows OS, and Macintosh OS X all have integrated support for 802.11 wireless NIC drivers. Almost all the major NIC vendors support 802.11b (and more are offering support for 802.11a integrated into the same wireless products). Since this book illustrates the vulnerabilities in WEP, we now look at how a laptop running Windows XP exemplifies using WEP authentication procedures in a typical wireless environment.

Windows XP supports the following types of features and functionality:

- **Automatic network detection and association**—Wireless NIC cards employ a logical algorithm to detect any available wireless network and associate with the best one in range.
- **Media sense**—This feature is used to determine when a WLAN NIC has roamed from one access point to another. As a result, it may require that you reauthenticate yourself and employ other types of configuration changes that must be set properly so that you don't compromise the security of your wireless network.
- **Network location support**—This functionality allows Windows applications to be notified as soon as the computer roams through the wireless network. Programs also have the power to update their settings automatically with respect to the changing parameters of the current network settings.
- **Power mode support**—Wireless NICs are automatically told when the power coming from the laptop device is from an AC adapter or the battery. This information makes it possible to conserve energy when necessary and shut down the system (or put it to sleep) to save power and extend the operating life of the laptop computer when it is in use for mobile applications.

Enhancing Mobile Security

Microsoft, dominant in the mobile operating system environment, is working to develop security solutions for its products. The company has formed a new division called the Security Business Unit to find out how to expand security opportunities.

Microsoft is constantly creating proposals for the next generation of ciphers that are based on the advanced encryption standard (AES) and are applicable for both 802.11 networks and IP security (IPsec).

Remote Users

Remote users constitute a predominant portion of Microsoft's user base. Windows XP and Windows 2000 Server offer the capability for other

Windows users to log into the "remote desktop" of another machine from a laptop computer. What is most interesting is that while either a Windows 2000 Server or Windows XP Professional computer can offer remote desktop services, almost any other Windows version can install a client and log into the remote machine from either a dial-up line or a high-speed Internet connection.

Microsoft is working to bridge the gap between the Macintosh and Windows environments. In so doing, Microsoft has created a virtual remote desktop client for any Macintosh computer running Mac OS X 10.x or above. This client allows the Macintosh to connect to any Windows XP Professional or Windows 2000 server running "Terminal Services," the service that allows remote desktop sharing.

In terms of security, it can easily constitute a vulnerability to the target server. The reason is that all communication takes place on Port 3389, so if you know the port you can attack the target machine with a variety of usernames and passwords in an attempt to breach security and gain access. Furthermore, if you know which machine you are hacking into, you might already know the account username. However, in most cases, a hacker will simply target the "administrator" account because that exists on every machine. The hacker need only keep hacking into the machine to find the password for this account.

Securing the remote connection Microsoft has officially recommended the use of VPN solutions for any remote type of connection. This means that both PPTP and L2TP VPNs provide strong security for any user attempting to perform business transactions across the "unsecured" Internet.

Conclusion: Evolving Laptop Security

The modern enterprise is constantly evolving, and the need for a secure laptop computer to access your wireless network resources is absolutely essential. In this chapter we have seen how WEP has limitations with respect to potential vulnerabilities in your security.

Security must involve a combination of solutions, regardless of whether you are using a Windows, Macintosh, or Lindows OS-enabled laptop computer. Windows XP, Mac OS X, and Lindows OS all have integrated support for 802.11 as well as other features and functionality to access wireless network resources.

Regardless of which OS platform you use, there is one common fact that must be observed when dealing with security issues on your wireless mobile devices—implement a solution above and beyond the integrated features present within your access point.

Realize that no device is secure out of the box. 802.11 has a number of safeguards that may not be enough to secure your system, but all OS platforms have the ability to utilize encryption and screen out computers not authorized to access your network. Your first objective is to enable the highest level of safeguards possible when configuring your laptop devices to access your wireless network. These settings may not be sufficient to protect your laptop from a determined hacker, but will make it harder to access network resources with any stolen equipment.

Finally, look at your laptop device with the eyes of a hacker. Know how to password-protect your computer so that nobody can even boot the device without knowing your personal password. Many laptop devices also include support for biometric devices that restrict access to computer functionality unless you authenticate yourself to the device with some personal information (fingerprint, retina scan, etc.). Use these devices to make it difficult, if not impossible, for anyone to know how to turn on your computer except for you and authorized people in your company.

If you take these simple steps to protect your equipment, you can save your wireless network from any hacking attempts. Create an access barrier at each level within your company—from laptop to access point and then (and only then) can you function in a realistically secure wireless networking environment with your laptop computer.

CHAPTER 16

Administrative Security

The most common error people make when it comes to wireless security is when administrators and/or users fail to change their default passwords, or create passwords based upon readily determined factors such as users' names, birth dates, and pet names.

This chapter explains common mistakes and shows how to administer wireless network security shares so people do not gain fraudulent access. We will examine the mechanisms by which hackers circumvent security so as to determine a path you can effectively follow to help you administer the effective lines of a "secure defense" for your wireless network.

Authentication Solutions

How do you administer better security? You add a number of very carefully tailored authentication solutions so that only authorized wireless network users can access your WLAN (Figure 16.1).

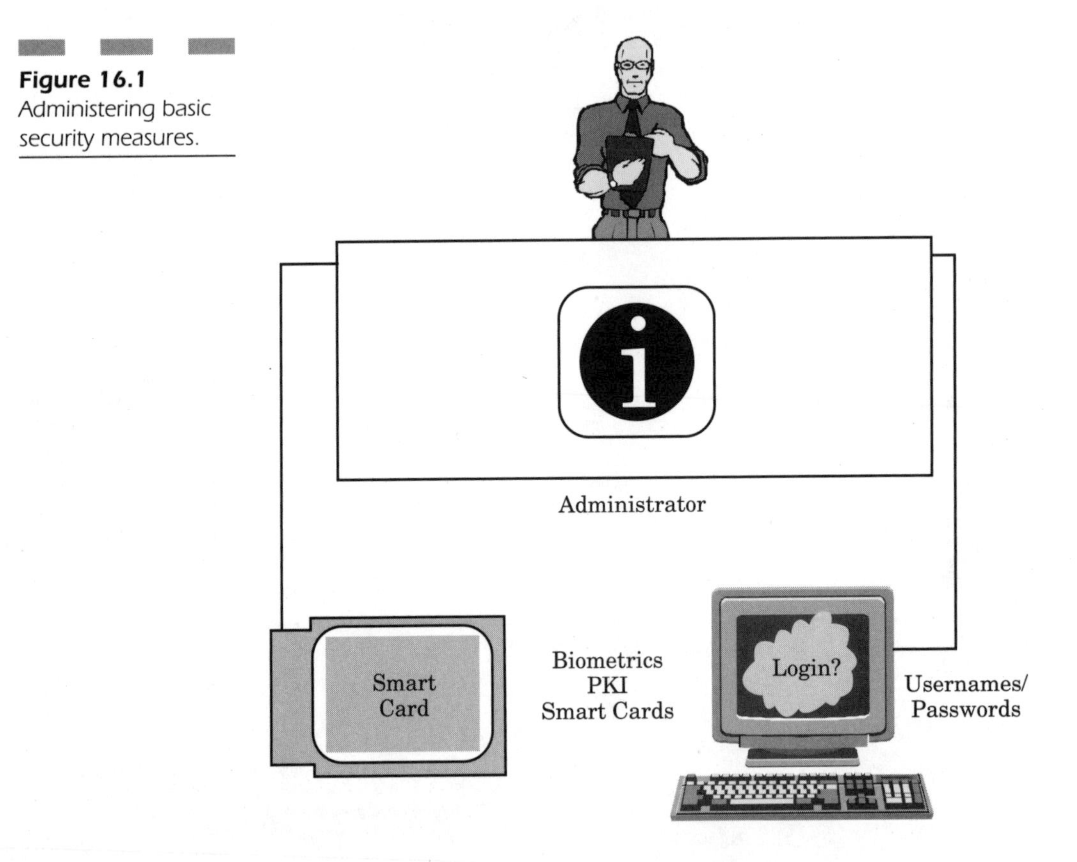

Figure 16.1 Administering basic security measures.

Authentication solutions are built primarily on creating more secure:

- Usernames/passwords
- Biometrics
- Smart cards
- PKI

Most solutions rely on a combination of these technologies to provide the most effective method of authentication. For example, when your only means of authentication relies on using just usernames/passwords, then it is extremely important to have a policy that designates some critical criteria:

- Minimum password length
- Specification of alpha or numeric characters in the password (i.e., Joe324Frog)
- Password should not contain either a name or dictionary term (in contrast to the example above, your username should be something like: J34D46Glop)
- Username/password expiration times (forces the user to change the password so it doesn't become stale or intercepted)
- Personal smart cards
- Biometric devices (so you know exactly who can access your network)
- Public key infrastructure (uses a criterion where only the sender can encrypt transmission data in a specific way, so that you *know* who is sending you information)

Passwords

You can also use more effective passwords for any parameters on your wireless networking devices or access point. However, it is important to note that the encryption scheme and other settings will then have the minimum protection available.

Building the Firewall

Why is the firewall important? If you consider networked information that resides on public network, then you realize all your information is virtually unprotected.

Now, extend the concept of unprotected information to the store of data on your wirelessly connected laptop. If a hacker were to gain wireless access to your wireless information, he could take data directly from your device as easily as accessing a network repository without any firewall. This is why it is generally very important for each wirelessly connected laptop to have a personal firewall protecting its information.

For example, Norton Antivirus is updated every year and can function on both the PC and Macintosh platforms. This program is highly recommended because its virus definitions are often updated automatically without any type of user intervention. These programs consume very little network bandwidth most of the time and are good at preventing hackers from sending any viruses to your machine.

Protecting resources on your machine from unauthorized access requires a slightly different program. Another component of the Norton Utilities is the Personal Firewall (part of the Internet Security suite of software). This program installs on your workstation and prevents other users from accessing your hard disk as a file server.

Another program is called Zone Alarm and enables a personal firewall that can be configured with various levels of security. You can prevent any incoming connections to your computer or just enable a light level of security. The ability to customize your protection allows you to access specific resources on the Internet that require greater access privileges.

Just like the Antivirus and Personal Firewall programs mentioned above, this product can be securely updated by the central server at the manufacturer's Web site to protect you against new threats that appear on an almost daily basis.

There is also the benefit that you can change your environmental settings on the fly, to screen out hackers trying to ping the ports of your firewall in an attempt to breach a suspected vulnerability.

In order to provide truly secure access, you can configure your firewall to accept only incoming connections using a virtual private network (VPN). The idea is to protect your internal resources by adding an extra layer of protection against any attack where a hacker expects to gain wireless access to the network.

Intrusion Detection Systems

The best method to allow you to make certain your wireless network is protected against intruders is to implement and carefully monitor an intrusion detection system (IDS), so that unauthorized users are caught trying to access your network. If a hacker does access your network, the IDS will send an emergency alert to the administrator of your network with the hope that he will catch the attack in progress, find the open vulnerability, and prevent the hacker from accessing the network in the future.

When dealing with a wireless network, the intrusion detection system you choose can be a host-based intrusion detection system (HIDS) or a network-based intrusion detection system (NIDS) (Figure 16.2).

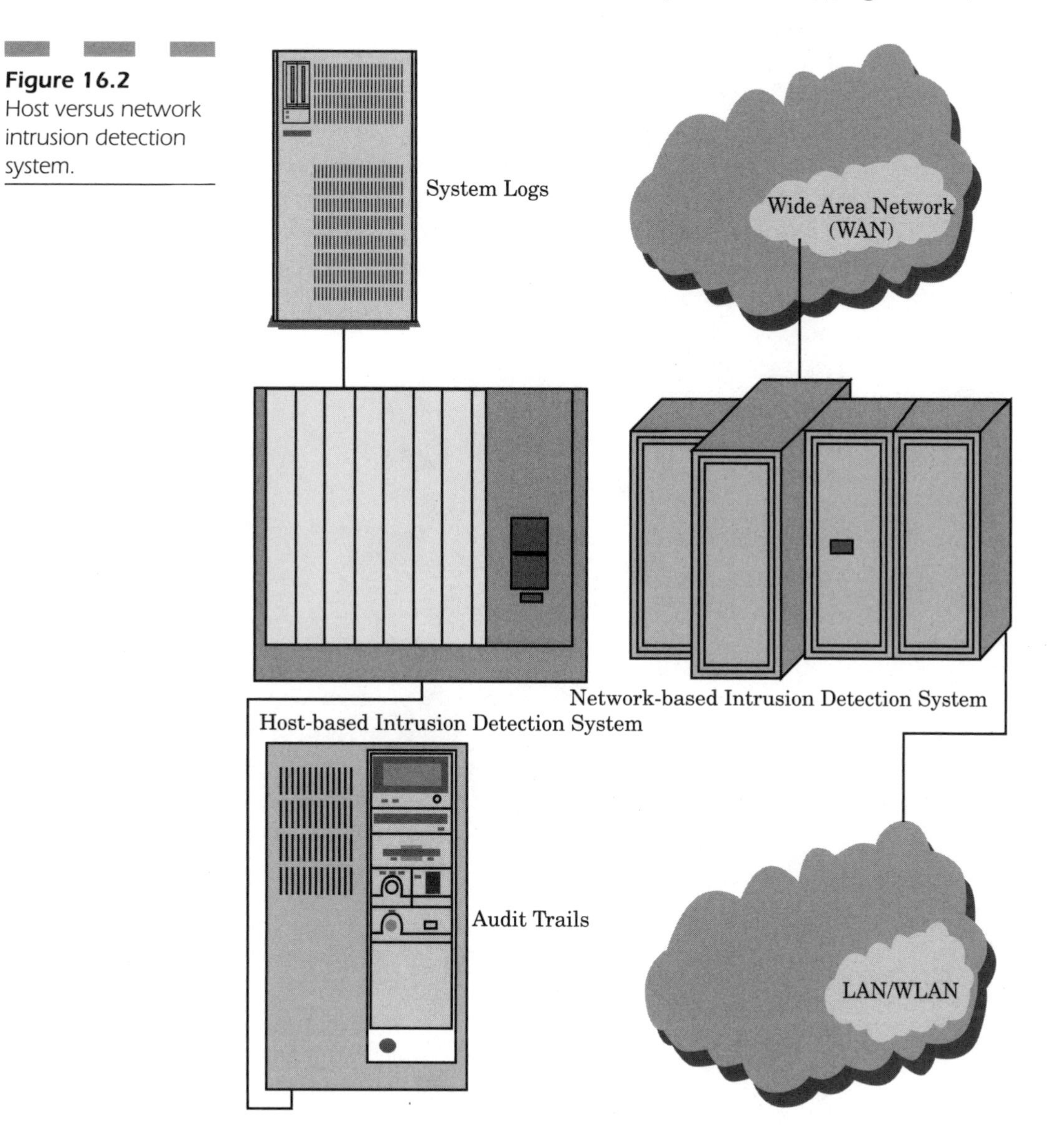

Figure 16.2 Host versus network intrusion detection system.

Host-based IDS

Host-based intrusion detection systems specifically look for vulnerable systems. They use a host-based agent that works on each server in order to monitor both the system logs and the audit trails for any activity that might indicate a hacker trying to breach your security.

Hacker behavior An intrusion detection system looks for specific behavior indicative of a hacker trying to breach your network (Figure 16.3). This type of activity will more than likely include:

- Modifying file permissions
- Multiple failed login attempts
- Excessive "after-hours" activity
- Failed access attempts on multiple accounts
- Spikes in activity (indicative of a program trying multiple login/password combinations)

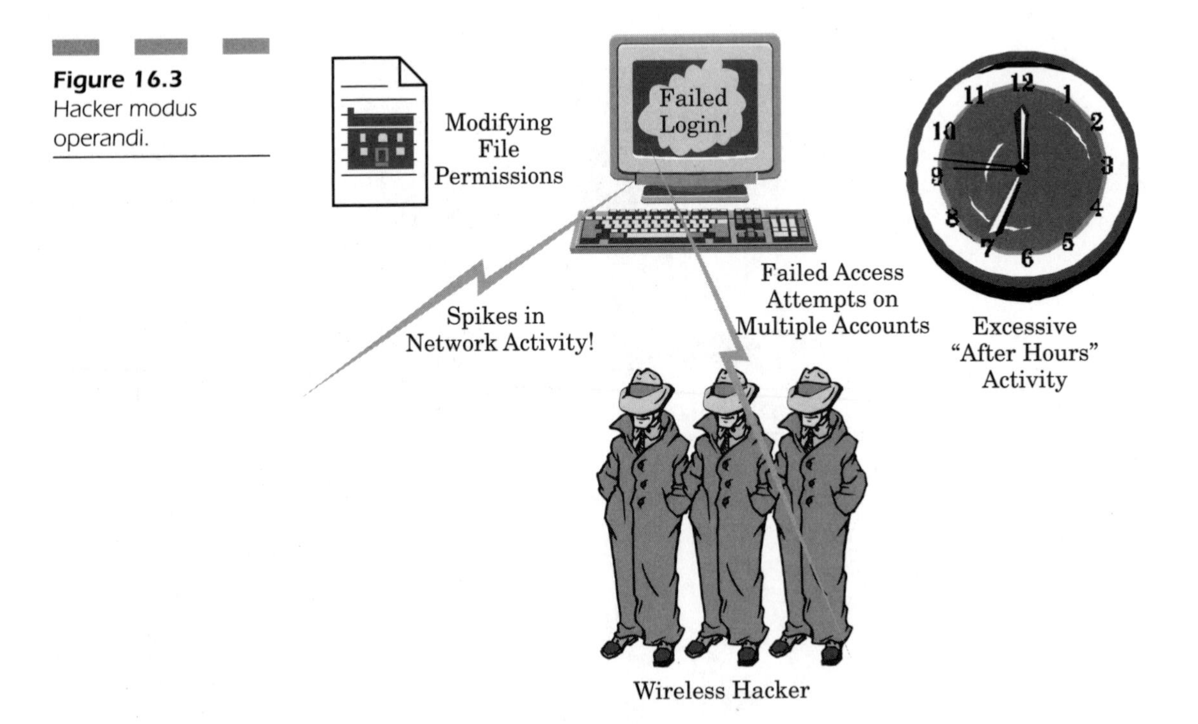

Figure 16.3 Hacker modus operandi.

A good host agent can analyze an attack in progress, determine from the log that a malicious event is happening, and immediately send an alert notifying the network administrator that a hacker attack is in progress. The only useful way to protect your systems is to know of an attack as soon as it occurs (preferably before) since information is the best weapon of defense.

Network-based IDS

The network-based intrusion detection system monitors both the LAN and WLAN in an effort to examine every single packet of traffic as it is transmitted across the network. The idea is to ensure that this traffic matches any known (preprogrammed) attack signature that might indicate a hacker type of attack.

The most common type of attack is the denial of service (DoS) attack, in which a hacker bombards the wireless network with so many packets that literally no other traffic can flow across the network. The idea is that if the hacker can't access any of the network resources, nobody can.

A good NIDS will understand this type of attack pattern and then summarily disconnect the network session from which these incoming packets originate. The IDS will also send an immediate alert to the administrator so that the administrator can take immediate action to prevent any damage.

Host IDS versus Network IDS

In general, the advantage of implementing a host intrusion detection system outweigh those of a network intrusion detection system, especially when it comes to dealing with encrypted transmissions. This is primarily because encryption protocols are more easily handled when dealing with either SSL or VPN connections through the firewall.

HIDS can look at the data transmission *after* it is deciphered. NIDS cannot, because the IDS agent itself sits on the component. This means that the encrypted data channel is sent right along through the network without having first been checked for attack patterns.

Why Have an IDS?

There is a very important reason why you *need* an intrusion detection system—it gives you an essential layer of security that you must have in order to keep wireless hackers from gaining access to your network without your knowledge.

In fact, even end-users are strongly urged to implement personal intrusion detection systems into their wireless workstations because it is an important layer of security. An administrator can view the logs your IDS generates in order to track and prevent hackers from gaining access to either the network or your laptop hard drive.

There are different types of intrusion detection systems:

- Computer decision (the computer determines if the alert warrants an e-mail alert to the administrator)
- Real live people (an IDS center makes the determination if your computer is being hacked—even if it is a slow hack over a period of days or weeks)

The Computer as the Decision Maker

Many intrusion detection systems are founded on the philosophy that the computer is smart enough to recognize an attack when it is coming in. In order for that to be true, an experienced security expert must "predefine" classic attack patterns that the computer can recognize and flag as real attacks. This is similar to creating attack strategies in chess; however, as with any computer, the strategy can be defeated by a real human being who uses a unique strategy to win the game or attack the host system, as the case may be.

When the computer makes decisions, it assigns each hacking strategy into a specific category that specifies exactly what type of attack occurs. Each attack is then classified into a severity event—measuring the severity of the attack on a scale from 1 (the least problematic) to 5 (melt down). Most systems are configured to send an alert to the administrator when an event of level 3 or greater occurs. Under this type of system, the computer must have accurate data regarding each attack. The database of "attack signatures" should be updated on a frequent basis by the vendor, in much the same way as virus signatures are updated when a new virus is discovered.

Some computers are now using what is called "fuzzy logic," which can dynamically identify an incoming attack and measure it loosely against the attack signatures in the database. Hacker attacks are not straightforward; in fact most of them involve diverse strategies that do not match up "exactly" with preprogrammed attack scenarios. The computer can use fuzzy logic to approximate incoming wireless network activity to determine if security is being breached. If the activity does appear suspicious, the IDS will then generate an e-mail to alert the administrator to the suspect activity. All these actions occur quickly, since no human intervention is needed to identify problematic network attacks; this gives the administrator greater time to catch a hack "in progress" and take the necessary steps to stop the attack or backtrack it to its source, for the potential prosecution of the malicious party.

A company called Intrusion.com builds systems like the one described above. In the majority of cases, hacker activity does not happen all at one time. Many hackers attempt to access your systems a little each day. Sometimes these probing activities last for days or even weeks. When a hacker probes your network only a little each day, it is done with the intent to stay below the radar screen of your IDS. The hacker has no desire to be caught, and he knows that only spikes of activity indicate a possible attack.

These computer IDSs are, however, prepared for low-level hacker activities. The systems keep a log for a period of approximately 28 days looking for discernible patterns. This is done on the philosophy that a hacker will "make his move" within a month of initiating attacks on your systems. With such a large time frame, the computer has a good foundation to draw upon in order to make decisions about potential threats to your computer network.

Real Live People

The other type of IDS doesn't rely on fuzzy logic or predefined attack signatures—instead it relies on *people*! Yes, they still do exist when it comes to evaluating potential problems with your network systems, and in many ways they have an edge over the computer being the decision maker.

Counterpane is a good example of a company that builds an IDS that installs in the corporate environment and then sends information about network activity (logs) back to an evaluation center for trained personnel to determine, over a period of time, if you are experiencing any type of hacking activity. Although this type of situation is not nearly as quick as the computer-generated alert example above, it does eliminate false positives when the computer keeps telling you that you are under a hack attack when you really aren't.

The idea is that a computer-generated system can be only so accurate when it comes to knowing how to identify hacking attempts against your networks and other systems. When you have a real-live person looking at your logs on a continuing basis, you have the security and knowledge that a person is the best judge possible of how many access attempts are really taking place. If someone is indeed trying to break into your systems, then a service set up specifically to identify possible attacks is the best judge.

The whole idea is to make it possible to perceive that a bigger attack is coming down the line. Your best defense is having an expert who can inform you of possible problems when it really counts.

In a setup like this, the IDS company installs a machine inside your network which sends reports and information through a secure, encrypted channel back to the home office, where analysts review the data. The biggest worry most companies have is whether or not the IDS machine poses a possible risk—a hacker that could gain entrance to the network through the very device designed to prevent breaches? The answer is that these servers are configured so that only authorized personnel can access limited information pertaining to access activity and logs. The IDS machines themselves do not have access to the mission-critical data flowing across the network and therefore should not normally constitute a security vulnerability if compromised.

Security Vulnerability Assessment

Administrative duties are often overwhelming when you have to worry about security for all your employees who are moving data across a wireless network. There are a number of wireless network analysis sniffer tools to help you determine the extent of your wireless network coverage. However, fraudulent access points designed to capture traffic or facilitate unauthorized access usually represent the most dangerous breed of hacker.

The best offense is a good defense, and you can effectively defend yourself against fraudulent access points by exercising extreme caution when implementing a WLAN in your corporate environment.

Consider one WLAN implementation that allows workers to use their wireless workstations anywhere within the limits of the corporate facilities. Prior to deployment, your security personnel (or third-party consultant) will execute a security risk assessment to determine what vulnerabilities exist within your proposed wireless infrastructure. The idea is to have your security experts (white hat hackers) try to exploit these vulnerabilities in an effort to determine your exact risk when running your WLAN and how any problems or security lapses will affect your organization.

The administrator now has the advantage of deploying your wireless infrastructure more effectively after assessing your risk and deciding if that risk is greater than the benefits offered by a WLAN. The benefit is that understanding these problems beforehand will allow you to reduce your overall risk before you implement your WLAN. This knowledge allows you to administer and utilize your wireless resources far more effectively, so that you can make certain you have the greatest possible level of security protection from the design to the deployment phases of your wireless network.

Risk Assessment

Once you are able to determine your level of vulnerability, you can determine your overall risk assessment and how best to direct the computer security in your organization to identify the countermeasures you should take to reduce your risk prior to implementation (Figure 16.4).

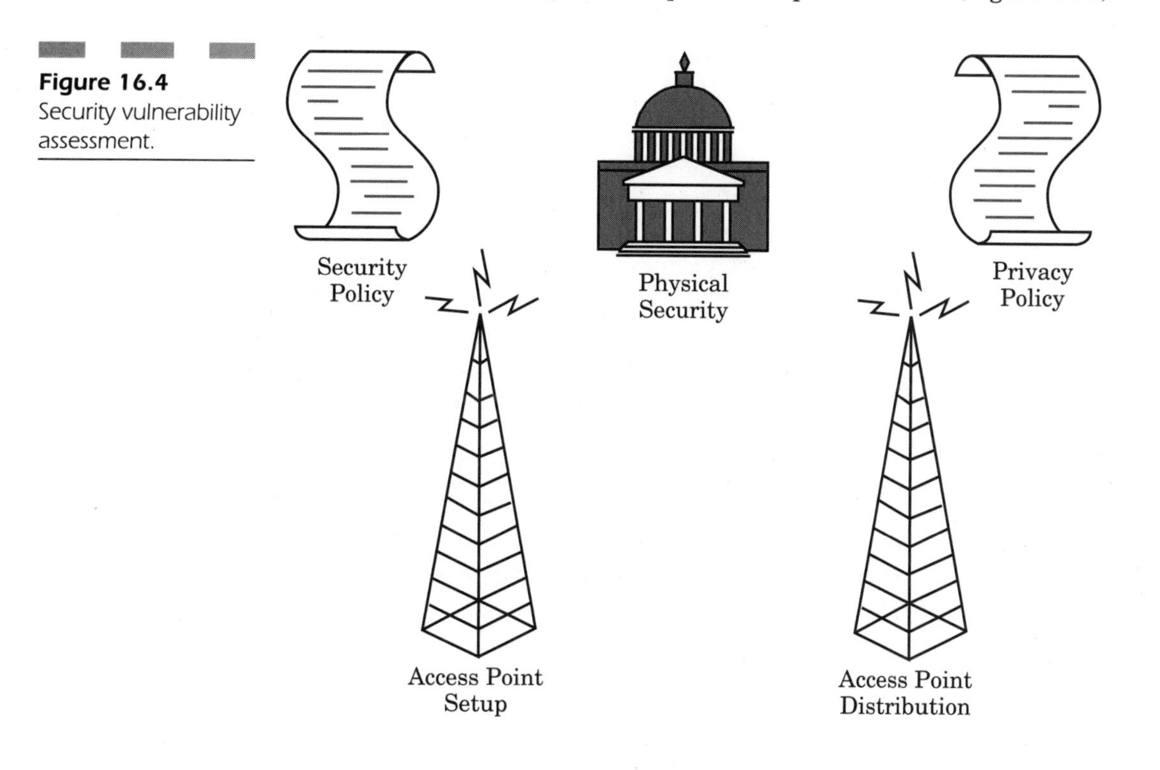

Figure 16.4 Security vulnerability assessment.

Five primary areas of security are important for any level of risk assessment. These include:

- Security policy
- Privacy policy
- Physical security
- Access point setup
- Access point distribution

The most serious vulnerability is a breach of physical security, which occurs when any unauthorized person not an employee of an organization is able to gain access to the corporate facilities. In order to make certain that only authorized employees and contractors enter your corporate facilities, you need to adopt, and make certain you continue to use, physical security safeguards such as:

- Biometric identification techniques
- Magnetic card badges
- Photo identification

You must also have a real-live security team (and this doesn't always bode well for contract security companies) who are actually part of your organization and know what to look for when screening individuals for admittance into your facilities.

The biggest problem that security guards face is hackers who use "social engineering" techniques to gain access into your corporate facilities. There are so many excuses and methods by which you can claim to enter a building—and almost any guard will feel duly pressed to allow hackers into the area under the legitimate belief that they need to be there based on what they said.

You ultimate objective is to make your security team understand how to make certain that your wireless network is *not* accessible from outside your corporate facilities. This means you must carefully examine each and every access point within your organization in an effort to realize exactly how you can prevent eavesdropping that may result from unforeseen network vulnerabilities.

Site security is often assured through survey assessments that make certain you have placed all your access points in the least accessible locations within your organization. The reason for careful placement of your access points is to make certain nobody can alter or modify your configuration settings.

As an administrator, you should physically map where and when users access your network. Just remember that there are a number of high-gain antennas that can pick up wireless signals at great distances. This makes it even easier for a hacker to eavesdrop on your WLAN. However, you can mitigate this risk simply by using your wireless network independent of the main firewall in your organization. You should also require that any incoming connection traffic use a VPN to encrypt the data channel so that even if the signal is intercepted, it won't make sense to anyone.

Risk is sometimes difficult to predict; this is why the precautions listed here will help you mitigate your risk while you can still take advantage of your WLAN. Be aware that many new hacker tools come into circulation all the time. For example, new encryption breaking programs have risen to the level where "script kiddies" (any would-be hacker) can just launch a program to monitor your wireless transmissions in the hope of determining any vulnerabilities that exist within your WEP encryption algorithm.

Since WLANs pose a risk if not maintained properly, your best defense is to enable the following critical safeguards (Figure 16.5):

- Random WEP encryption keys
- Access control lists
- Virtual private networks (within your wireless connectivity)

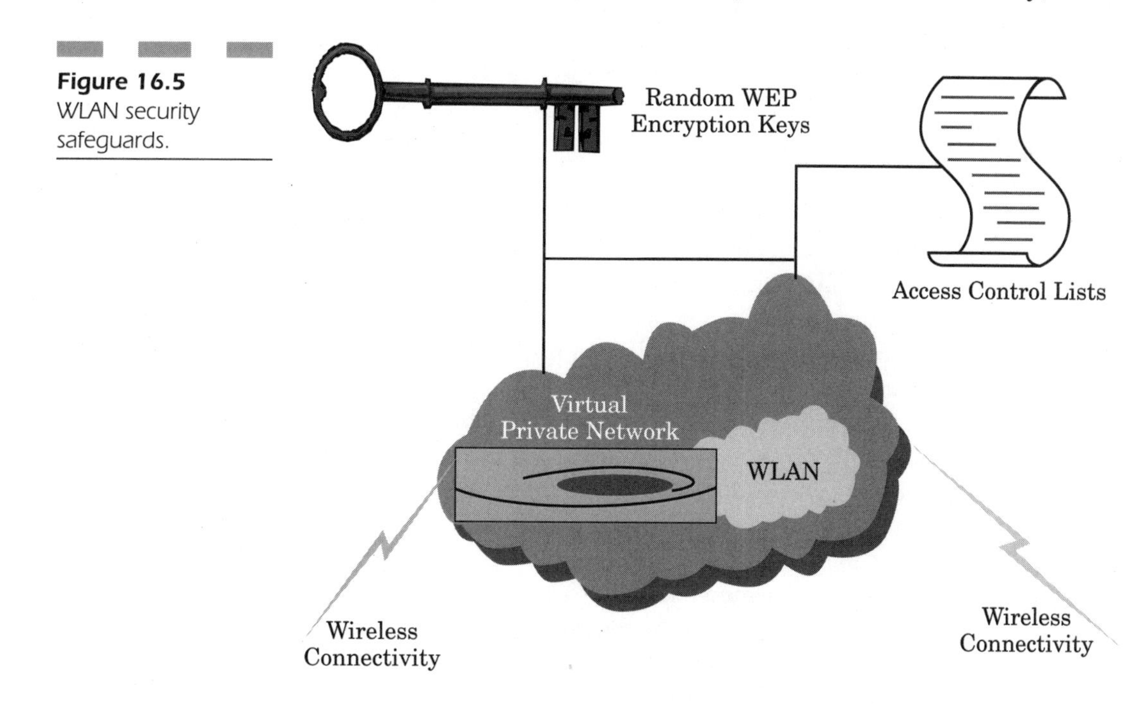

Figure 16.5 WLAN security safeguards.

Defense programs are becoming more and more sophisticated as they offer enhanced security solutions that extend throughout both the wired and wireless sections of your enterprise.

Conclusion: Best Defense Is a Good Offense!

There are a number of steps you can take to administer your security in the most effective manner possible. You can use the steps outlined here as a reference guide to implement the necessary safeguards to ensure that your wireless network is secure at all times.

As we have discussed in this chapter, there are multiple "layers" to your security solution. These layers often include physical security, access levels, and most important, the administrative types of security. The administrator is the "key" or cornerstone of your entire wireless network. If anyone is going to try to breach your network, the administrator will be the first line of defense in preventing your information and network infrastructure from being corrupted.

Protecting your network involves the adoption of good physical security. This entails preventing unauthorized users from any access. Adopting a personal identification system for every employee and contractor within your organization is important to achieving the control you need.

That control also extends to the Web-based configuration for your access points. These devices are designed to be very easy to configure. Unfortunately, that ease of use can very easily translate into a security breach when someone comes into contact with the access point. A hacker can easily access a password-unprotected resource and alter the settings to allow unrestricted access into your intranet.

Sometimes the smallest and least thought of access control barrier is enough to buy you time to protect your company. For example, how good are your password rules? Do you have an alphanumeric password assigned to every member of your team before they acquire network access? Did you make certain there are no words from the dictionary in the password? This simple precaution would make you less vulnerable to a hacker using an automated "dictionary" attack, where every word from the dictionary is sent to your login prompt in order to gain access. Are your employees forced to change their password every few months to make certain that the information never becomes "stale" and therefore susceptible to discovery by a hacker? Do you have a rule that states that *nobody* is permitted to share a password with any other user, *no matter what the reason*?

The most common mistake administrators unfamiliar with wireless networks make is not turning on the inherent WEP encryption capabilities. Often, you will need more security than simple encryption, but I can't stress enough how highly I recommend using the highest-available encryption, presently 128 bit. The NIC cards that support 128-bit encryption (on average) only cost about $10 more than the regular wireless NIC cards. This expense more than justifies itself by making it that much harder for a hacker to breach the security of your network.

One of the biggest security vulnerabilities is that most administrators fail to realize that access points enable an "open system" right out of the box! Most hackers just wait for people to enable an open system so that they can come along and directly connect the network using DHCP, and no one is the wiser. Access point devices support ACLs that are configured to screen out any wireless NIC card whose unique MAC address has not been previous entered into its configuration access settings by the administrator. This very simple step does a world of good in preventing a hacker from roaming onto your network without your knowledge. This essential protection scheme *must* be employed as the most basic level of protection to ensure hackers don't gain access to your mission-critical internal network resources.

Another step you can take is to change the default SSID for your wireless network and make certain you don't allow just anyone to roam on your network or pick up your SSID just by eavesdropping when the network broadcasts this piece of information. Many network administrations feel they are secure as long as nobody knows their network SSID. Nothing could be further from the truth; this is the easiest way to hack into the network, because the SSID can be determined by a little social engineering or just by finding the field blank as it is in most wireless network cards.

The most important test is to have a security team come in and perform a study of your network in an attempt to determine items such as the best placement of your access points, and to identify if your signals are vulnerable to attack from a hacker trying to roam onto your network, eavesdrop, or simply disrupt the wireless transmission by making your entire WLAN useless to any user (similar to a DoS attack).

Personal firewalls and VPN transmissions are a good way to make certain that when a connection does take place from the outside, it is at least structured to enter the protected internal network through the designated ports in the firewall; that transmission should also be encrypted using a VPN so that nobody can eavesdrop on your signal. Firewalls are not only for the server, but for the wireless workstation too. Processing power in laptop computers, for example, has become as powerful as that on any server in many cases. These machines can easily be exploited by hackers attempting to turn the wireless laptop into a file server. Information from your internal network can be stolen just as easily from the laptop as it can from the mainframe itself. This is why inexpensive personal firewalls are always a good idea on both ends of your wireless connections.

Finally, you should *at all times* establish a wireless security policy. Make certain that when mobile workers travel, they password-protect all their access connections; sometimes a simple password can be required before the device is even allowed to boot up! Establish your access policy and make certain users follow it. Simple steps will help you make certain that you can effectively administer your WLAN so that you make it enormously difficult for hackers to penetrate your defenses. Although security is never 100 percent, forewarning of an attack, preventing gaping security holes, and ensuring that users follow a predefined policy and procedure before accessing mission-critical internal network resources are all that is needed to make certain that you can maintain security and justify the safe and secure deployment of a beneficial wireless network that will meet your information needs effectively and efficiently for many years to come.

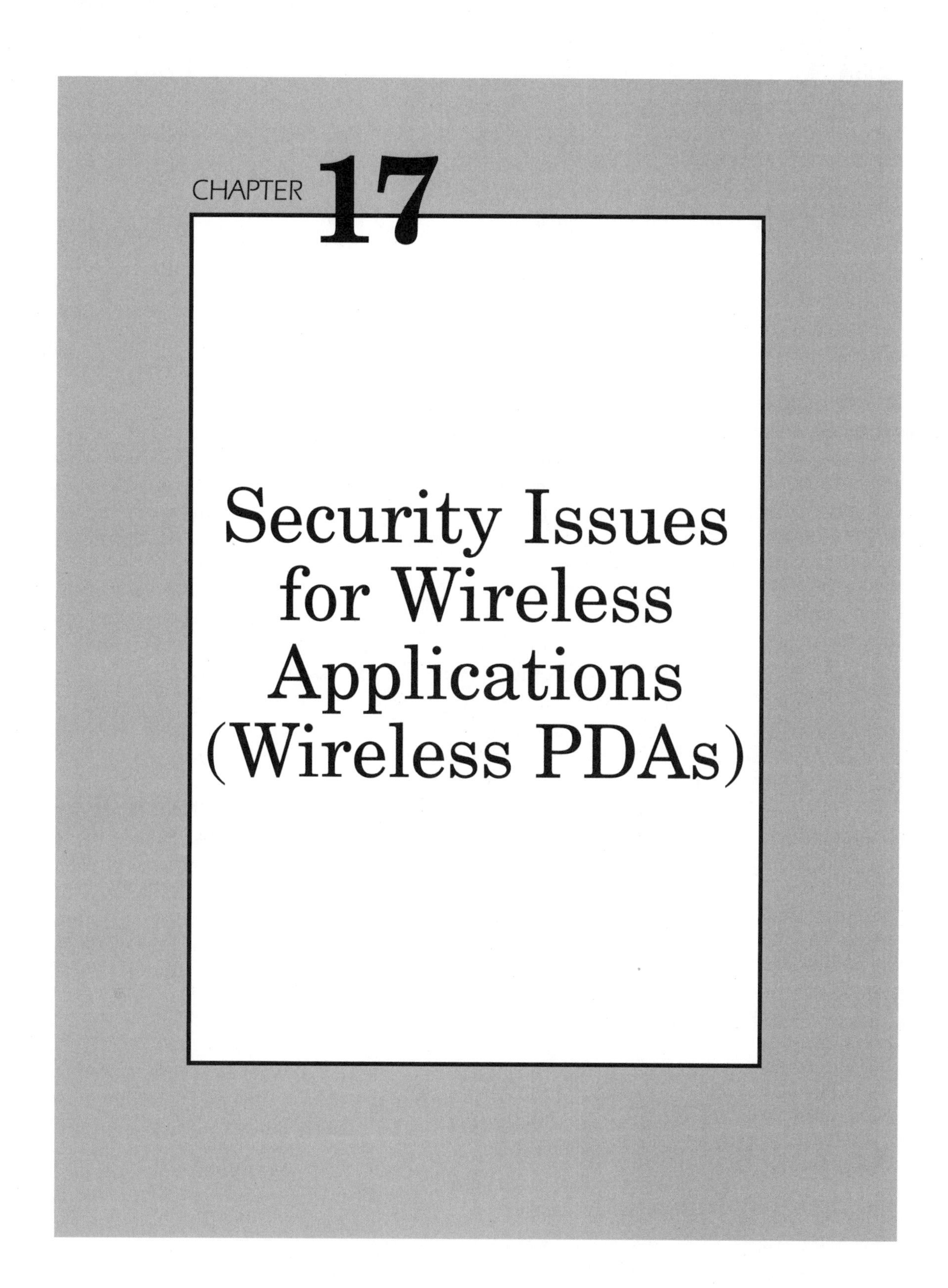

CHAPTER 17

Security Issues for Wireless Applications (Wireless PDAs)

This chapter describes how the evolution of wireless applications will generate an entirely new set of security issues, making users prone to over-the-air hacker attacks. As wireless devices are gaining more momentum from Palm-, PocketPC-, and PDA-enabled telephones, 802.11b is commonly built into these devices, thus making it all too easy for hackers to compromise your data stream, access protected documents, and destroy mission-critical systems.

Protecting Information

How many of us have handheld computers? Their evolution has grown significantly, to encompass a great deal of information, databases, and confidential documents. Today, most handheld computers can support an endless supply of flash memory that can hold hundreds of confidential documents.

Both Microsoft PocketPC and Palm OS mobile computers have applications that support reading and writing in Microsoft Word, Excel, and even PowerPoint applications. The amount of confidential information in a given office document could compromise entire projects.

The newest Palm devices include support for the 802.11b protocol built right into the device itself. These devices are designed to enable users to roam from one wireless network to the next while maintaining full access to the corporate network.

PocketPC 2002 has compact flash slots, while its latest OS offering provides integrated support for 802.11b. These mobile devices are leaner versions having the same type of functionality as Windows XP in many ways. The evolution of this operating system allows it to roam seamlessly from one wireless network to another.

PDA Data

The question that you really need to answer is: how secure is all that data residing on your PDA? PDAs are now capable of containing such confidential information (Figure 17.1). This includes:

- Mission-critical work data
- Personal information

- Contact lists
- Financial information
- Network passwords

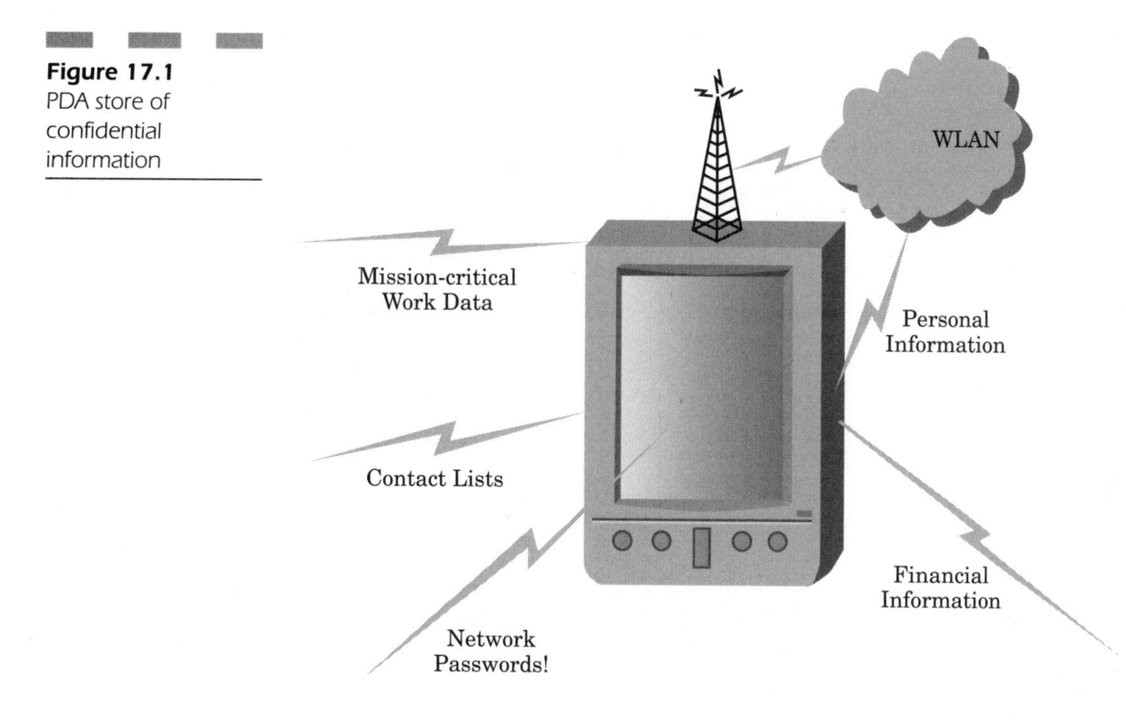

Figure 17.1 PDA store of confidential information

How many important and confidential business contacts exist on your handheld device? There is little or no protection on these devices to stop someone from stealing that information by either physically acquiring access to the device or connecting to it through synchronized operations with your wireless network.

Seeking Security

Many individuals and organizations are just starting to realize how vulnerable mobile PDA devices actually are. It is often difficult to find built-in basic security functionality that works transparently in any wireless PDA. Managing these functions from one central interface is difficult. For one thing, accessing these devices is often unprotected by any password. Second, there is no barrier to prevent any wireless hacking exploits against these devices.

Security Functionality

There are a number of critical functions that define the vulnerabilities within both the Palm and PocketPC platforms. If you are going to secure your PDA device, you have to consider all the areas in which your device connects to your network.

Access Control

Access control must be mandatory for all your devices. If you are unable to establish and stick to an enforceable policy, you are leaving yourself vulnerable to attacks by any hacker.

HotSync

HotSync operations are wonderful for keeping your PDA up-to-date, but they also leave all the information on your device open to attack. Most sync operations can now occur through either a standard Ethernet-enabled docking cradle or a wireless network connection. The problem with these types of sync operations is that the computer *never* asks you for any authentication information. Unfortunately, this means that anyone can gain physical access or wireless network access to your mobile device. Without any means of authentication, all the information on your device is left completely vulnerable to anyone interested in getting your confidential contact information.

Infrared

How many times have you wirelessly transferred a file or a program from your Palm computer to another? Most people perform this action on a regular basis and never give it a second thought! But have you ever considered how much in the way of information, programs, and contacts can easily be compromised by anyone interested in acquiring this information?

It is imperative that you make certain that not just anybody can pick up your handheld device and copy information from its memory through your wireless infrared interface. The best security mechanism designed to protect you from this type of information attack is to add the safeguard of authentication to your infrared device transfer capabilities. In this way you can make certain that infrared capabilities on your device are disabled unless you specifically authorize them to function (Figure 17.2).

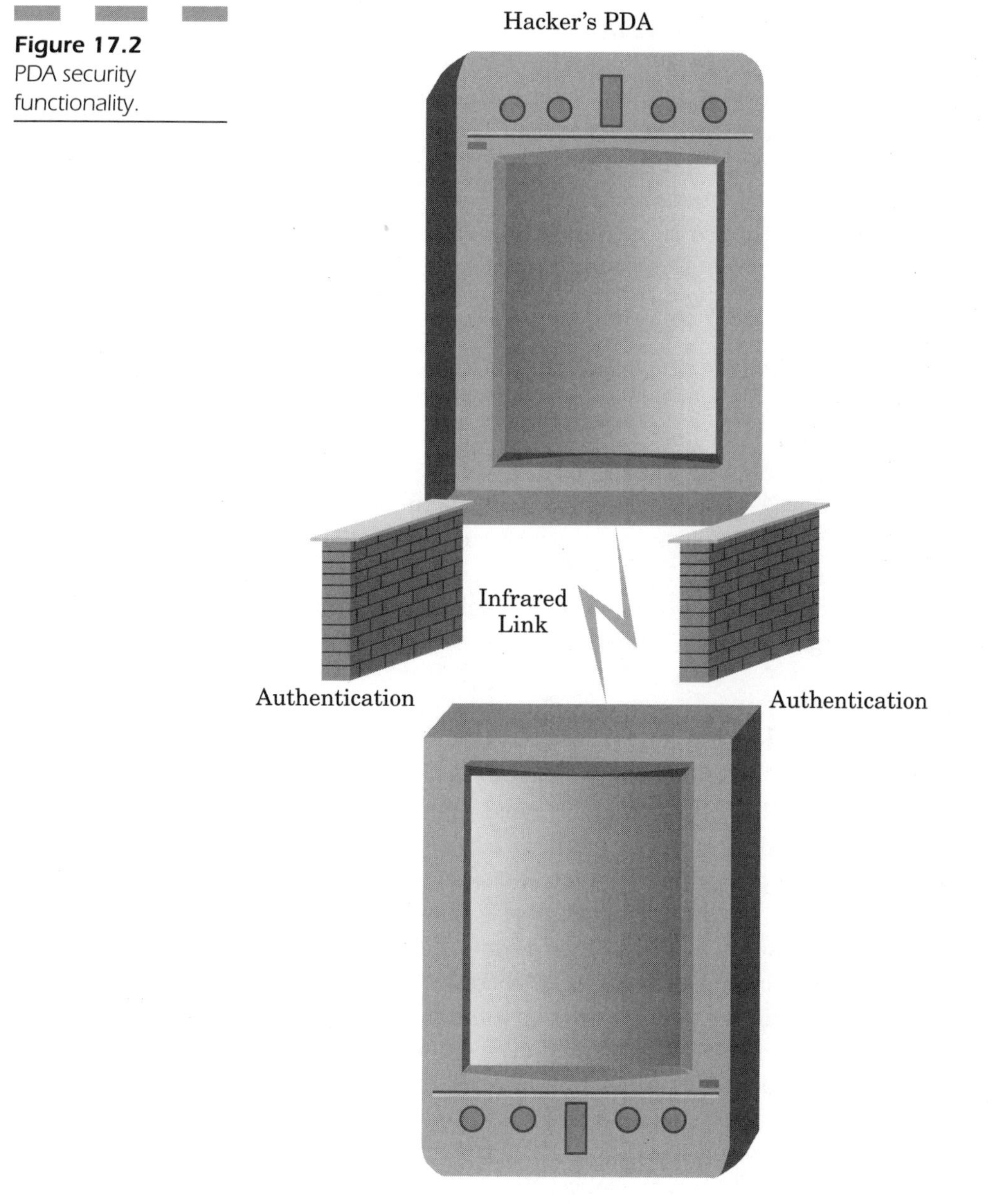

Figure 17.2 PDA security functionality.

Building an Effective Mobile Security Policy

Mobile PDA devices are inherently insecure, a fact easily recognizable from the lack of any realistic security standard. An effective method of ensuring mobile device security is to attempt to build an enforceable mandatory access control system to make certain that only authorized users are able to access your PDA and its confidential contents.

The best way to ensure that security is maintained is to make it centrally controlled by the administration within your organization, as opposed to the individual user. If a device is stolen or compromised, then it is that much more difficult for a hacker to gain access to the device or collect information or passwords.

Protecting Mobile Resources

When you protect your mobile device, it is important for it to ask you immediately for a password before any confidential content is accessible. Wireless operations that retrieve confidential data from your device are vulnerable unless you can lock out any HotSync and IR communication. The user who initiates this functionality must be successfully authenticated to the PDA. Lack of one defining authentication process activates all the features locked out by security protocols.

Wireless Connectivity

When the PDA is first turned on, the user should automatically be authenticated before using wireless connectivity or functionality. The minute a user gains physical access to your wireless device, he can easily find your network passwords, network settings, and IP settings. All this information is enough to give a hacker sufficient ability to hack at leisure into your wireless network.

Mobile devices provide carte blanche access into your corporate network. A hacker who wanted to find a way into your network would use the mobile device as a conduit between a workstation and your wireless network. This is a simple method of gaining access to protected resources.

Your wireless network should normally have safeguards that prevent any unauthorized wireless NIC from accessing your network. The unique MAC address of your portable device is already on your access control list, so it is a simple matter to use its inherent connection protocols to circumvent your security. Your mobile device already knows how to connect to your wireless network since it knows all your settings. If you do have any authentication mechanism, it is a safe bet that the PDA's user has already preprogrammed all the necessary usernames and passwords into the device. A hacker can take the mobile device and connect to any wireless resource in your entire corporate network (Figure 17.3).

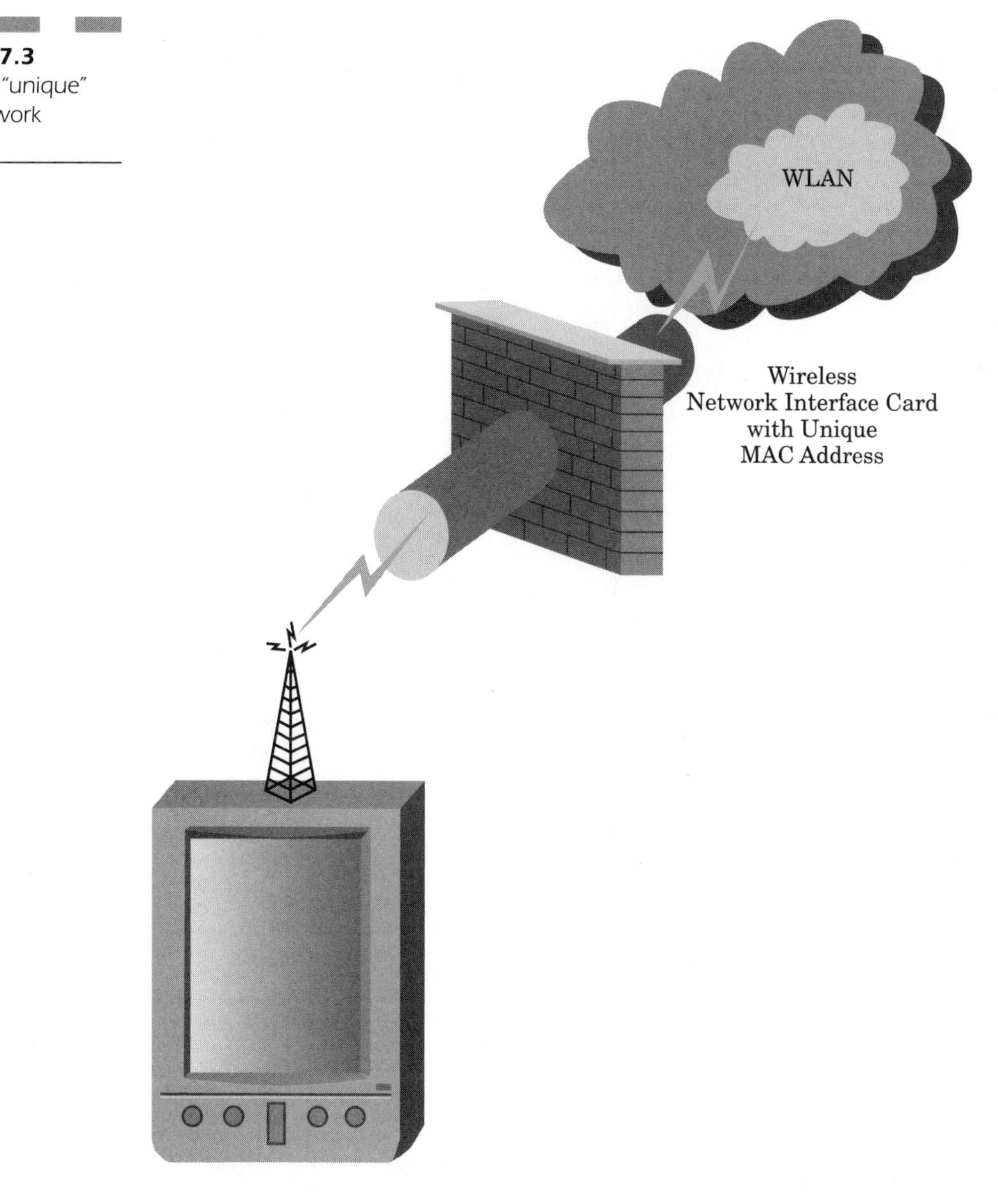

Figure 17.3 Spoofing "unique" MAC network address.

HotSync Security

Most PDA users never really take into consideration how vulnerable their mobile device is when left in the sync cradle at the desk or workstation. It is a simple matter for someone to come right into your office or cubicle, press the HotSync button, access your workstation, and gain access to all your essential contact and application information. The security vulnerability here is that all of this can be easily done without the user's having any knowledge that the unit is being compromised!

The answer to this problem is to institute an authentication mechanism that prevents anyone but the intended owner from using the HotSync operation at any given time. This simple precaution is enough to prevent any unauthorized person from gaining access to your names, addresses, applications, network settings, and passwords, all of which can compromise the integrity of your corporate internal or wireless network.

Infrared Authentication

It is easy for a hacker to use the HotSync operation to steal your data, network settings, and passwords. It is just as easy (if not easier) to steal information from your PDA by only having physical access to it for a short time. The infrared feature of all PDA devices is a convenient method of sharing contact from one PDA user to another; you simply point and click. In fact, it is so easy to transfer information that a hacker could use your PDA to download all your confidential contact information, network settings, and passwords directly to another device without your knowledge. This information can then be used to mount an attack against your corporate network.

An even more interesting example of how a hacker could really misuse your PDA is to download all your information, and then upload a rogue computer program, called a "Trojan Horse," into your handheld device. This could easily erase all your information, corrupt your data, make your PDA completely unusable, or even infect the corporate network with a virus as soon as the user reconnects to the wireless network with his handheld device.

With all these problems apparent from the lack of handheld device security, the best means of defense is a good offense. Authentication mechanisms are only now being developed for handheld devices. These mechanisms allow you to prevent anyone from initiating IR communication without first authenticating to the device. Unless you know the proper usernames and password (a password than *cannot* be entered into static memory), it would not be possible to activate any IR transfer functionality from the PDA.

The only problem is getting users to understand that this functionality exists and to *use it*! Security like this can provide a wonderful means of protection, but if it is not implemented or deployed, it becomes useless. Safeguard your mobile device so that your handheld won't be the cause of any malicious code getting into your corporate network and bypassing the firewall—because your PDA is already "inside" the firewall. Simple measures likes the ones described here are often enough to protect device and data security.

Establishing a Security Policy

When it comes to the security within your organization, you create a security *and* a privacy policy to protect your networked resources; why is it any different to develop a security policy to protect your PDA devices too?

All companies need to create a security policy that details the specific requirements and methods needed by a secure mobile security infrastructure. The wireless security policy you create is a set of rules for all the wireless resources your PDA utilizes. It is important for your security policy to define how PDAs are handled, what type of protection and authentication should be offered, and a set of rules for enforcing those standards at all times.

Your security policy also dictates exactly the type and how to configure the authentication settings for your PDA. These settings will control how each user is able to access the device. If a PIN is required to start the device, your security policy will determine the length of the PIN, what symbols are shown (instead of the actual letters or numbers) when typed into the device, and how many minutes should elapse with no user input or activity before the device automatically locks up to prevent someone else from picking up the device and compromising its internal data.

Flexibility is an important element in any security policy because you will undoubtedly need to change these settings later as your policy elements change. Your environment will change with respect to the security needs of the IT world around you. These changes will reflect methods by which you can strengthen security to deal with new threats and unforeseen vulnerabilities (Figure 17.4). Any wireless network is vulnerable, by definition. Your signals are in open "air space" and can be compromised, given enough time and interest on the part of a hacker.

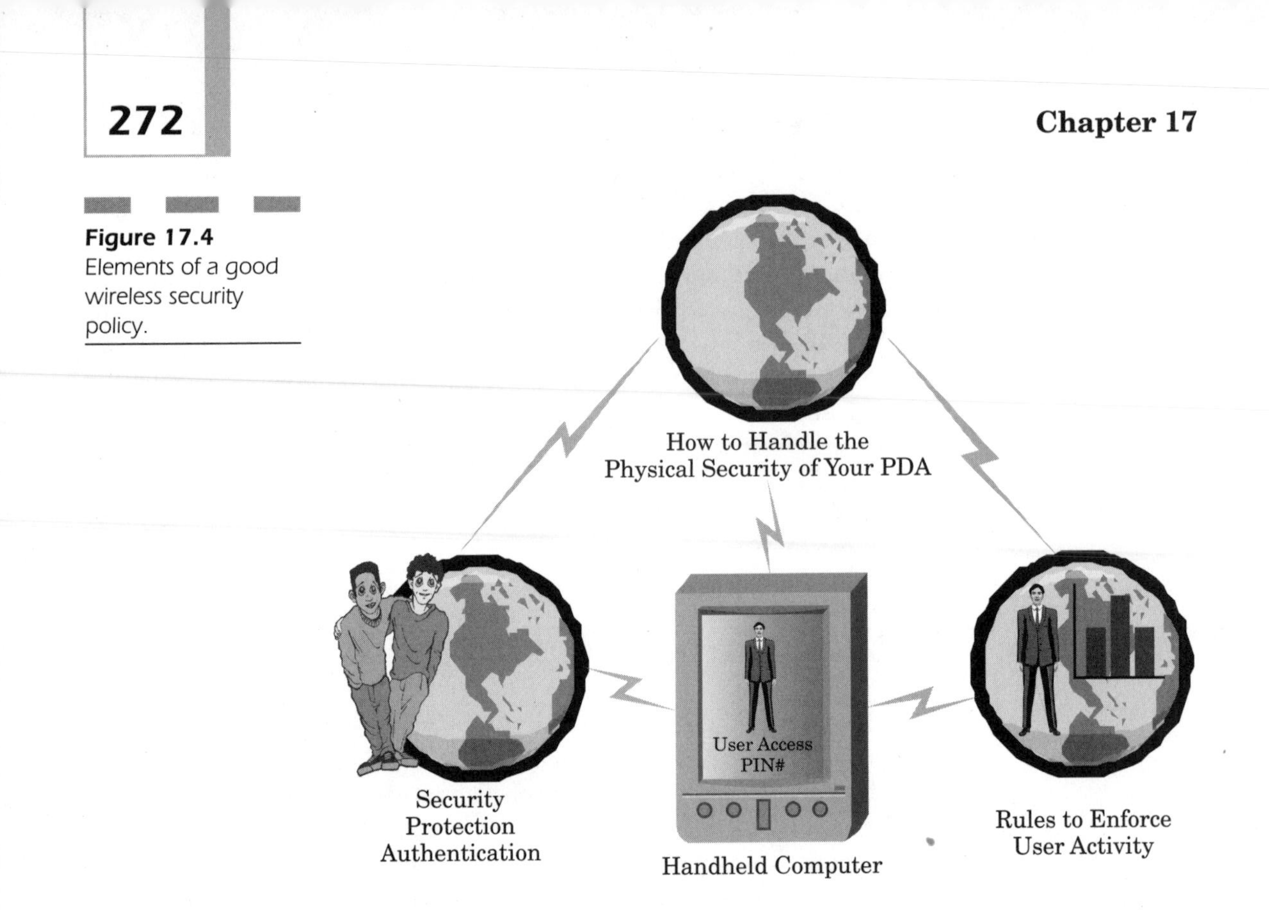

Figure 17.4 Elements of a good wireless security policy.

Privacy Concerns

Just as security is elemental to your ultimate success in protecting your network, privacy is important to protect the confidentiality of your information and the information of your customers. New privacy regulations are being developed all the time, and these trends will likely increase due to the vulnerabilities that 9/11 has shown the world.

Why PDAs Require Privacy

Privacy regulations are important for wireless PDA users. In all industry sectors, PDAs are essential devices that keep everyone connected wirelessly to the master network. For example, in the healthcare industry alone, it is essential for the doctor to have a PDA at his or her disposal to check drug interactions and send wireless prescriptions to the pharmacy directly from a patient's bedside.

The nature of these devices delivers information wirelessly from a master database to the PDA in the course of normal daily operations. Additionally, important information is entered into the PDA and transmitted wirelessly back to the main network server to allow the processing of confidential information that may contain names, addresses, and personal information.

Maintaining Access Control

How do you effectively make certain that you are doing everything possible to maintain the security of your PDA device? One good method involves building on the access control configuration options defined earlier in this chapter. It is not enough to establish a PIN to lock down your device on startup or lock it again if not used for a certain period of time; you need to configure a way to shut down the device after a specific number of unsuccessful attempts to access your device. Maintaining access control prevents unauthorized users from accessing your PDA's internal data. After too many attempts, the device should have a safeguard that allows it to lock down so that only the owner can reactivate the device.

When the user regains access to his PDA, he or she can securely reset the PIN and choose a new PIN to gain normal access to the device. This type of setup allows for a maximum level of security so that no unauthorized user is able to continue trying different codes in attempt to gain fraudulent access to the PDA.

Data Encryption

Another ounce of prevention will keep any unauthorized user from hacking directly into the data files of the device to try to take by force the information contained within your PDA. The best method of making certain that nobody can force information out of your device is to employ an encryption algorithm that scrambles the internal device information such that if anyone does gain directory access to your PDA (which is possible through a number of directory-browsing Hot Sync programs), the information would be useless without the proper encryption key.

SecurID

A good means of protecting information on your Palm device is afforded by RSA's SecurID. This product is a piece of authentication software installed on your Palm computer and used "positively" to identify network and system users prior to their gaining access to confidential resources. This product works almost exactly as standard RSA SecurID software authentication applications. SecurID is used in combination with the RSA ACE/server network security software.

It is important to note that RSA SecurID for the Palm creates a random, one-time use access code that changes every minute. This frequently changing value makes it exceedingly difficult (if not impossible) for a hacker to figure out how to guess this value and gain fraudulent access into either your mobile Palm device or your wireless network.

The advantage of integrating RSA SecurID authentication into your Palm gives you the power to have "crack-proof" security in one simple method. It also eliminates the need for the user to carry a separate hardware authenticator, which is normally necessary for other network access requirements, including networks that cannot be accessed unless you use a virtual private network tunnel from a public to a private network where a hardware device is required to initiate the proper encryption scheme.

Intranet Access with Your PDA

In today's geographically dispersed corporate world, mobile staff must have immediate access to updated company information by being constantly plugged into the corporate network. PDAs give employees that flexibility and freedom to do business while on the move, but having wireless access to all the resources on your corporate intranet can be a serious security risk because the nature of an intranet is private. An intranet is tucked away from the public version of a company's Internet Web site. Intranets provide employees with special access to corporate databases, sensitive applications, and remote services that provide confidential information on clientele.

In most corporate intranets, authentication with a username and password is required to gain access. From there, an authorized user can view practically any Web page, obtain private information, access confidential directories to find addresses of clients or other employees, and access an order system to add or view customer information for items, including product availability and inventory. Data about regions can easily be mined, and this information can be valuable to your competition.

How Hackers Fit into the Equation

Hackers look for open conduits into your intranet in an attempt to acquire information on your protected resources. They love wireless networks because WLANs represent a method of gaining access that a corporate wired LAN does not provide directly. With modified transmission sites, it may be possible to try to break into a LAN from the next building or even down the street. The point is that a determined hacker, given enough time and ambition, will find a way to either eavesdrop on your WLAN traffic or log into your network.

If the hacker happens to have a stolen PDA, the process of finding a way into your corporate intranet is much easier. One needs only to break through the defense of the PDA (which is not terribly hard to do) and use that information to mount an attack against your intranet. As it has become essential to do business with wireless messaging applications and productivity tools, wireless connectivity constitutes a necessity for mobile employees.

Security Concerns

Mobile wireless devices have the same problems with the wired equivalent protocol (WEP) in 802.11b as does any wireless workstation. Security is lacking in WEP-based security keys, especially when the values are kept the same for a period of time. This means that even a PDA can give up its secrets for accessing the network because hackers need only capture a few packets to descramble the rest of the security code that permits access into your network.

Understanding the similarities and differences when dealing with security in your wireless environment is important, because it can help you deal with the vulnerabilities in your wired network. When a hacker accesses internal resources over your wireless connection, your wired Ethernetwork is vulnerable to attack at any server as well as any access point.

PDAs as Diagnostic Tools

Wireless PDAs are not just potential units that can compromise your wireless security; they can also be used as potent diagnostic tools to test the validity of the security on your network.

Security vulnerability assessments are often performed by using a PDA as a testing system that can simulate attacks on access points within range of a WLAN in an effort to find potential cracks. These tools can generate reports that allow you to create fixes to resolve gaps in your security network architecture.

PocketDOS

One way that a PDA can be an essential diagnostic tool is to convert its operating system to Linux. PocketDOS (http://www.pocketDOS.com) has created a product that allows you to take your handheld or PocketPC and emulate DOS, Linux, and a host of other operating systems. The advantage of this tool is that you can use the convenience of a small device to empower it to run hacking tools to penetrate the defense of your WLAN and determine your security gaps before they present a problem for you.

Handheld computers (PDAs) for the most part use either Palm OS or Windows CE. The PocketDOS application works under all versions of Windows CE (including the newest PocketPC 2002 OS). With respect to this application, Windows CE has extended the definition of the personal digital assistant to encompass a series of commercial applications.

Windows CE has evolved to the point where there are a number of wireless network interface cards that enable these devices to log into a network seamlessly. PocketPC 2002 offers integrated support for Wi-Fi connectivity as part of its core OS offering.

PC emulation Even though Windows CE devices do not have the Intel x86 processors you would normally find in a desktop or laptop computer, they do have the capability to emulate a PC by using PocketDOS. This application set allows you to use your handheld device to run executable PC programs in a variety of differently emulated OS environments.

The two primary types of operating systems are DOS and Linux. There are a variety of tools that only run in these operating environments, and they can be ported to your handheld device. This program gives you the ability to have a small emulation environment to run different hacking and analysis tools.

PocketDOS specifications PocketDOS is an IBM PC/XT emulator that works on WindowsCE to emulate an 80186 processor. You are then able to run a majority of applications created for a PC-compatible computer running MS-DOS or Linux.

The flexibility of this program allows different ROM-DOS images to be stored on your handheld computer. In this way, you can selectively boot into DOS 6.22 or Linux upon command. The objective is to be able to run as many operating systems as you need, in order to emulate all the functionality you require.

This type of functionality is exceedingly useful, since it gives you the ability to create an entirely new set of functions through different operating system platforms and applications. All these features help you create a small, but complete, computing platform with your handheld device.

Wireless Service Providers

Wireless PDA devices have the capability of offering a highly portable solution to doing business almost anywhere. PDAs can take advantage of wireless network service providers to acquire wireless service through carriers including GoAmerica, Sprint, and AT&T.

GoAmerica Communications

GoAmerica offers a more targeted product for PDA users. The "Mobile Office" allows users to access the secure business applications shown in Figure 17.5:

- Enterprise resource planning (ERP)
- Customer relationship management (CRM)
- Sales force automation (SFT)
- Mission-critical databases

A service provider called Go.Web compresses and encrypts data in an effort to optimize it for use on all types of wireless PDAs and data networks.

Enterprise customers can access wireless data services on every major wireless data network, using speeds ranging as high as 128 Kbps.

SprintPCS

SprintPCS also offers wireless Web for business users who need to have access to their corporate intranet from anywhere, using its nationwide

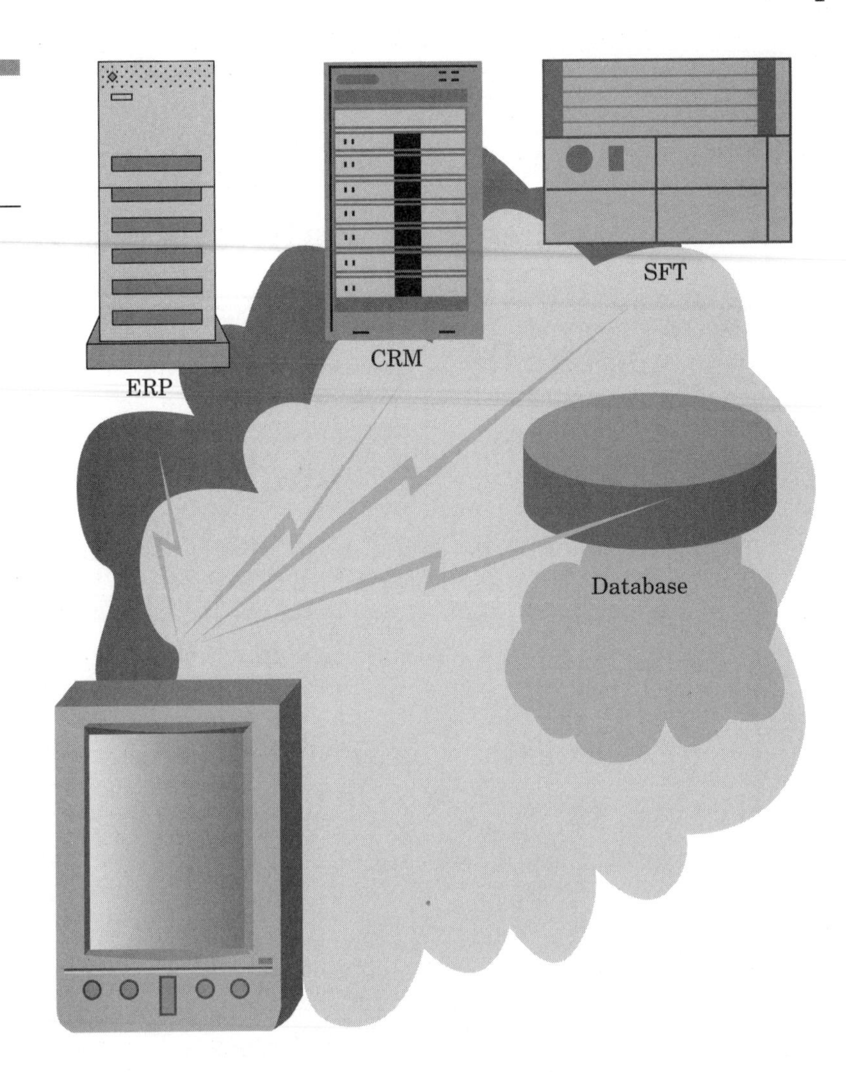

Figure 17.5 Mobile access to secure business applications.

network The standard connection speeds are equivalent to what you would expect with a modem, but Sprint is building out a higher-speed infrastructure. This type of connection allows your PDA or laptop to have access to any corporate resource from practically anywhere.

AT&T Wireless IP Network

AT&T offers a wireless IP network that provides speeds equivalent to that of a modem. The wireless connection they offer is encrypted for the purpose of securing the transmission of your information. Mobile employees can access information from a wireless network in most areas of the country.

Conclusion: Mobile Wireless Computing

Wireless PDA devices have really proliferated in the IT community. The applications of these devices are essential to doing business today. In order to effectively access the wealth of information in corporate databases, wireless connectivity from your PDA is a must. However, security is a big concern for these types of applications.

Every day you hear about new wireless exploits that hackers are creating to circumvent security measures and access protected information on a corporate intranet. Any compromise in access will undoubtedly reduce your ability to do business.

The goal of this chapter has been to describe all the possible vulnerabilities that exist for Palm and PocketPC devices. If you understand the problems that can result from these devices, you can learn to add precautions to protect your ability to access networked resources.

An ounce of prevention is often enough to help you prevent handheld device security problems. Make certain that *your* handheld device is protected as described in this chapter, so that you can be certain you are armed to defend yourself against hackers attempting to compromise the security of your mobile device.

CHAPTER 18

The Future of Wi-Fi Security?

As wireless LAN technology is adapted for use by more and more mission-critical applications, the threat of being hacked and losing data through security breaches increases. This chapter describes the evolution of wireless technology and determines how the threat of compromise can be dealt with on a daily business level.

Privacy Regulations

In order better to understand the changing face of the world with respect to privacy concerns, consider the following regulations, which define how privacy has affected mobile devices that carry the confidential information now frequently transmitted across wireless networks.

Patriot Act, 2001 (USPA)

Mandates the establishment of due-diligence mechanisms to detect and report money laundering transactions. This establishes new privileges of law enforcement and U.S. special services to intercept and obstruct terrorism. Among many other provisions, the Act mandates the establishment of due-diligence mechanisms to detect and report money laundering transactions through private banking accounts and correspondent accounts.

Graham-Leach-Billey (GLB) Act, 2001

Protects the privacy of personal nonpublic information shared by financial institutions with third parties. GLB core privacy provisions address financial institution disclosure policies regarding consumer information, consumer "opt-out rights," enforcement mechanisms, timing for implementation of regulations promulgated pursuant to GLB, and preservation of state jurisdiction.

Fair Credit Reporting Act, 1970, 1996 (FCRA)

Designed to promote accuracy, fairness, and privacy of information in the files of every "consumer reporting agency" (CRA). Most CRAs are credit bureaus that gather and sell information about consumers; for example bills are paid on time or if an individual has filed bankruptcy—to creditors, employers, landlords, and other businesses.

Children's Online Privacy Protection Act of 1998 (COPPA)

Legislates parental consent for use of information about children. This legislation makes it is unlawful for an operator of a Web site or online service directed to children, or any operator who has actual knowledge that it is collecting personal information from a child, to collect personal information from a child in a manner that does not, among other provisions, include parental consent.

Health Insurance Portability and Accountability Act (HIPPA) [August 21, 1996]

Enacted as part of a broad Congressional attempt at incremental healthcare reform. This requires the United States Department of Health and Human Services (DHHS) to develop standards and requirements for the maintenance and transmission of health information that identifies individual patients. Healthcare providers and health plans are required to create privacy-conscious business practices, which include the requirement that only the minimum amount of health information necessary is disclosed. In addition, business practices should ensure the internal protection of medical records, employee privacy training and education, creation of a mechanism for addressing patient privacy complaints, and the designation of a privacy officer.

Pervasive Computing

Wi-Fi is evolving to the point where it will encompass much more than computers and PDAs; it will involve almost any device enabled to access a network. This means that literally any device could be able to use Wi-Fi.

Take the example of a regular soda machine. A vendor could stand to make a great deal more money on the price of a soda can if he could raise the price during very hot days when people would pay more for a can, and then lower the price during cool days when people would only buy a can if it were less expensive.

One way that Wi-Fi is evolving is that a suggestion came from IBM that a "smart" soda machine could be developed that has a Wi-Fi connection to the vendor. This link would update the price per can of soda with respect to the weather that day. On hot days, the machine would charge more, whereas on cooler days the price would drop. This is an effective example that describes how the evolution of Wi-Fi can affect the economics of even our most basic types of transactions.

Wireless Mobile Computing

Mobile computing does not simply extend to the wireless laptop or PDA, although most Wi-Fi implementations are set up to communicate to these devices. In fact, wireless mobile computing is already evolving into something much...smaller.

Already, we see Wi-Fi being used in high-speed information retrieval systems for cellular telephones. Mobile phones used to be able to send and receive short messages that displayed news and information about current events. E-mail and instant messaging play a large role for "smart phones" too.

Today, mobile phones are evolving to act as true wireless replacements for the modem. Higher-speed wireless network deployments allow a direct connection from the mobile phone into your laptop or handheld computer.

Higher speed is not only useful for data transmission, but mobile phones are now sending and receiving graphic images for full Web browsing as well as image transfers. As the speed of Wi-Fi networks increases, it will be possible to raise the rates of real-time video so that each mobile phone user can see the person he is talking to. At present, only a few frames per second are possible, but as wireless network bandwidth increases, real-time video will be possible in the not too distant future.

Evolving Security

As Wi-Fi evolves to offer greater levels of functionality, security must evolve even more quickly to combat the new vulnerabilities that continue to plague wireless networks.

Basic Encryption

The 802.11 standard deals primarily with wireless deployments connected to wired networks. The most prominent weakness of 802.11 is that most deployments do not use even the most basic level of encryption. Unencrypted networks invite hackers to eavesdrop on or log into your network regardless of the presence of any session authentication. Even the most basic level of encryption prevents your resources from being looted.

WEP

Wired equivalent privacy (WEP) represents the "minimal" level of encryption described in the previous section. WEP provides you with a basic level of protection, but unfortunately, there are several weaknesses in this encryption method, all of which are easily exploited by wireless hackers. As 802.11 evolves, greater levels of secure authentication and key management may help strengthen some aspects of WEP. However, the problems WEP has are mostly due to its lack of support for per-packet integrity protection.

The number of attacks against wireless networks is increasing in scope, so it is important to look for alternative methods of security for your wireless network, simply because current safeguards are not enough to ensure its integrity.

Protecting Access

One method that can help you protect your wireless resources is to deploy mandatory mutual authentication that requires that traffic not be sent until the user is authenticated to the server. Unfortunately, most implementations to not offer authentication options, so support will come from the evolution of access points with this capability.

Some of the early 802.11 implementations are not able to use the per-session keys derived to encrypt your data transmission. These implementations only encrypt data using multicast/broadcast default keys.

These types of implementations are vulnerable to a number of WEP attacks, especially if the default keys are not automatically changed frequently and do not follow any given pattern for such changes. The administrator of your deployment needs to use automated methods to change the default key to create secure management mechanisms that keep your wireless network protected against these types of attacks.

As hackers become more and more savvy, and acquire a wealth of automated hacking tools, your network security *must* evolve to deal with these constant threats.

Denial of Service Attacks

When a user logs off the access point, a hacker can potentially pick up the transmission frame in an effort to spoof his or her identity. This essentially "tricks" the access point into thinking that the user has not really logged off the wireless network. Thus, the hacker can gain access or flood the unit with packets to effectively deny service to all other users on the WLAN segment.

There are a number of packets that initiate and end the wireless network transmission. DoS attacks (Figure 18.1) can literally take place within any WLAN deployment. The problem is that vendors have yet to implement a good source of interference protection capabilities that will stop a malicious user from flooding the frequency so that transmissions become unstable. Access point devices need to evolve so that modified packets do not effectively deny service to legitimate WLAN users.

Evolving Standards

The evolution of WLAN standards has focused more on increasing the speed of transmission. As growth continues, security will constantly hamper the adoption of WLANs for fear of hackers gaining access. Wireless networks, by definition, are open and subject to attack. It is imperative to add stronger levels of encryption, managed together with authentication, to ensure that data transmissions are secure. When a hacker attempts to break into your wireless network, he or she looks for the path of least resistance. As wireless standards evolve to encompass greater speed, make certain that you add as many security measures as possible to make the job of hacking your wireless network that much harder.

Figure 18.1
DoS attacks.

Competing Standards

Data privacy will always be a concern, regardless of whether you are using 802.11, Bluetooth, or HomeRF. Most cellular phones and wireless laptops can have their information exposed to the world if two steps are not taken to prevent a hacker from finding your information (Figure 18.2):

1. Add a personal firewall to your device and make certain you upgrade it constantly, otherwise a new hacker attack will circumvent your security and allow someone to gain access to your protected information.
2. Encrypt your data channel. It is *so very important* that ensure that if a hacker does try to eavesdrop on your data communications, he is not able to understand or read your information. *Don't* rely on the built-in WEP encryption—*always* use a VPN to scramble your data, or risk its being deciphered!

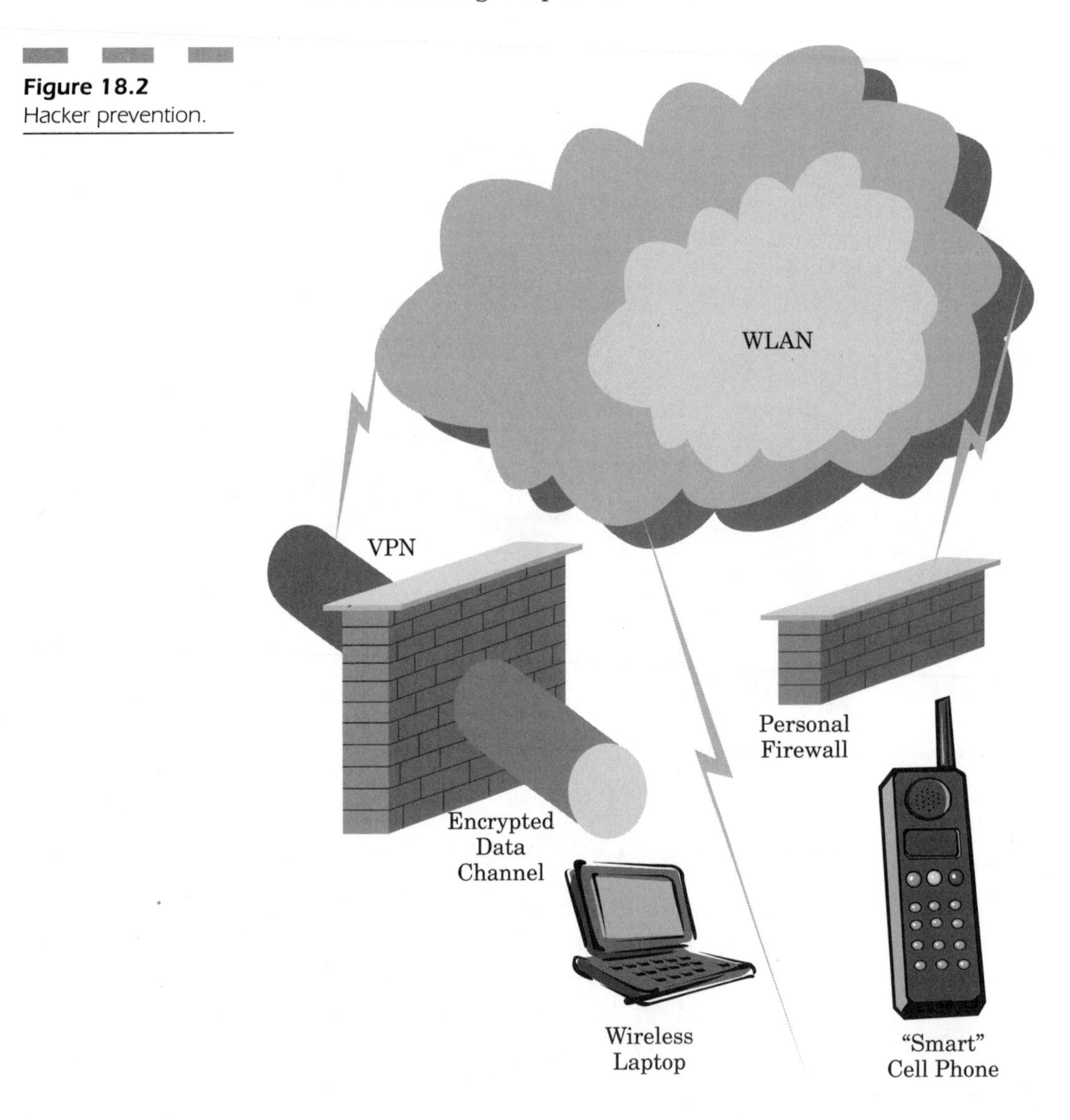

Figure 18.2
Hacker prevention.

Enhancing Your Wireless Security

Security can only improve (Figure 18.3) when:

1. The vendor of the wireless product implements greater safeguards (beyond what the specification has defined) to improve security.
2. The administrator in your company implements all these safeguards (and then some) to lock down user access to any wireless resource on your WLAN.

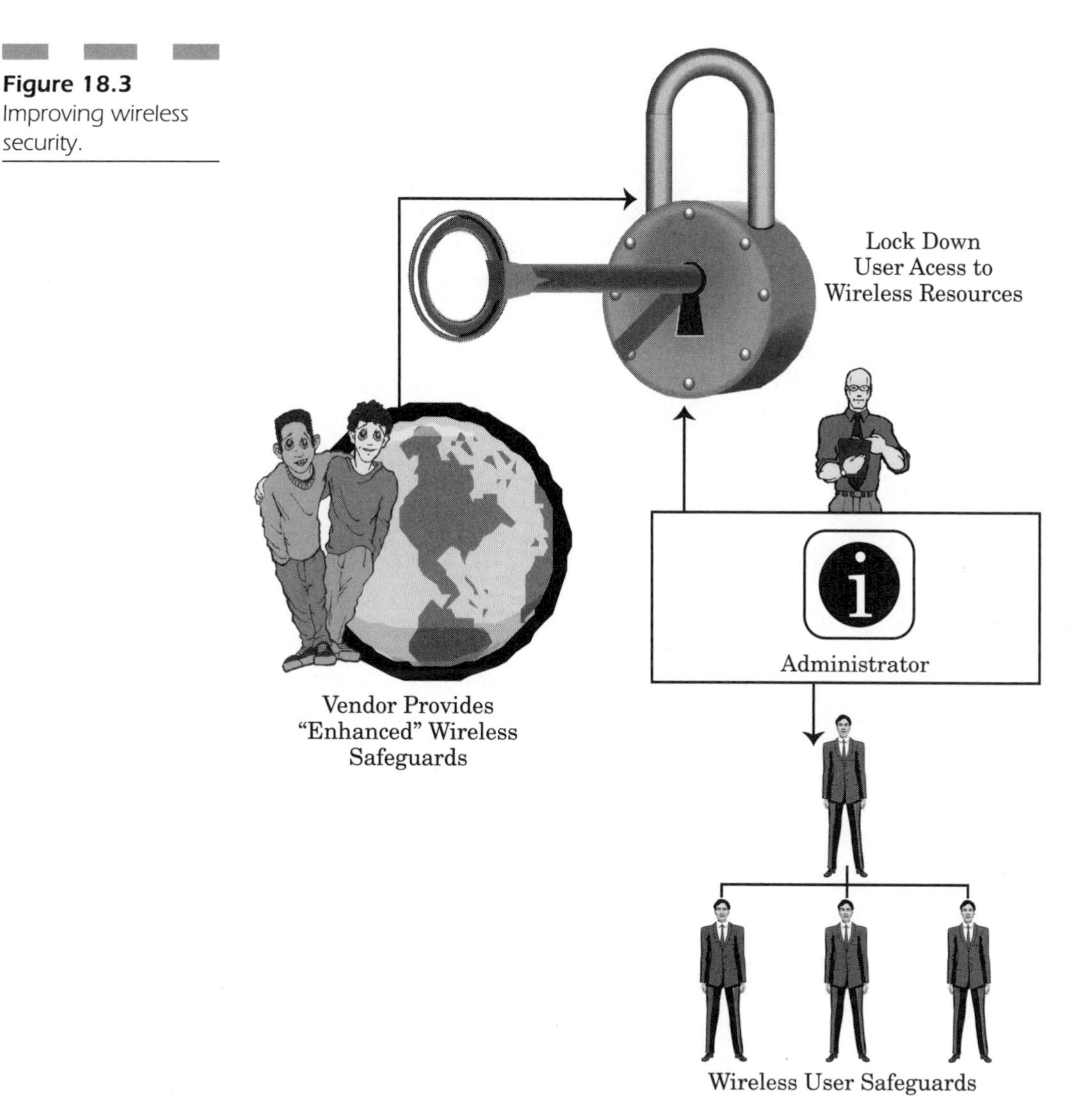

Figure 18.3 Improving wireless security.

Unauthorized access is a primary concern for any current Wi-Fi applications. Improving on security demands that the data stream be protected with encryption and that user access be protected through authentication.

WLANs may evolve to integrate PKI types of access barriers that allow selective access requiring only specific credentials. This type of access depends on granting a digital certificate to a user when he logs into your network. Such a certificate will only allow the user to access certain networked resources. This type of mechanism allows you to retain control over who accesses your network and what resources they can use.

Biometrics

Biometrics are a means of making certain that you can prove the user's identity before he even uses a wireless terminal to connect to your WLAN. Biometric devices make it possible to allow a user access based on some physical attribute (retina scan, fingerprint, or voice identification). A hacker would not normally be able to fake someone's personal attributes without being a spy with extensive resources.

Biometric devices are more cost effective today than they were years ago and can easily be integrated into laptop and other mobile devices. They provide a partial answer to protecting your network resources.

Assessing WLAN Strengths and Weaknesses

As your resources grow, it is important to have a security vulnerability assessment performed periodically as a "checkup" to make certain that your WLAN doesn't experience any holes in its security that could potentially damage your ability to host a wireless network (Figure 18.4).

Items to check include:

- Looking for improved ways to protect the integrity of your data transmission
- Isolating yourself from neighboring radio interference
- Maintaining your access control list, which determines how network resource access is controlled

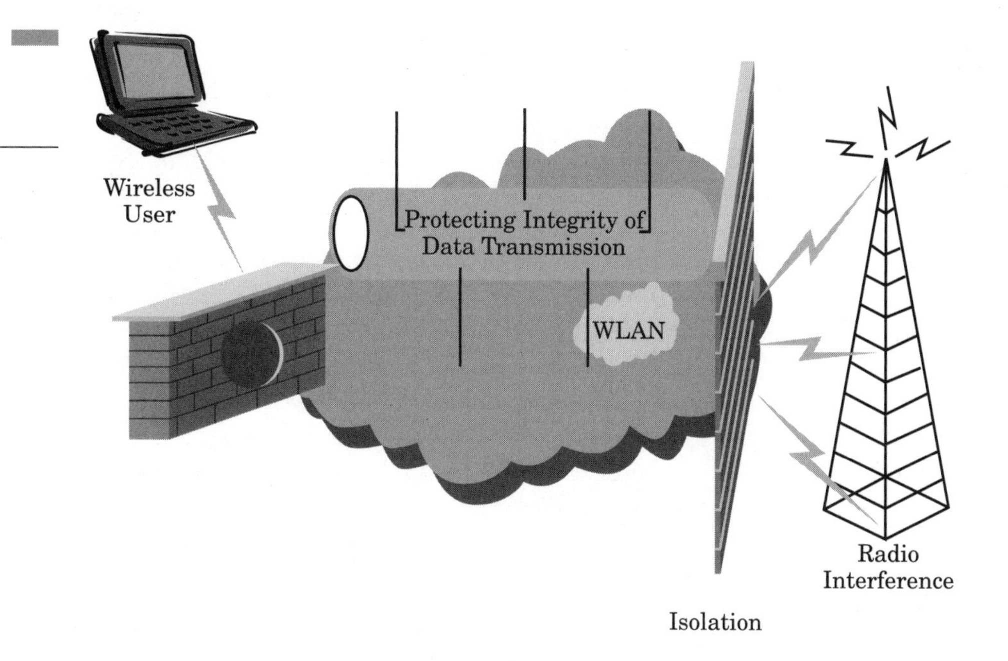

Figure 18.4 Protecting your WLAN.

Combining Future WLAN Technology

WLANs are constantly evolving to provide greater bandwidth for multimedia applications and still handle an encrypted data stream without degrading your throughput to the network. The most commonly deployed wireless networking standard today is 802.11b, but we are on the cusp of widespread deployment of 802.11a devices that are backward compatible with all 802.11b networking hardware.

Even mobile devices are evolving so that they can provide greater levels of computing power that fits literally in your palm. 802.11b is integrated directly into the Palm i705. Wireless communication is becoming a core function of handheld devices. However, it too suffers from all the inadequacies of an insecure infrastructure. Therefore all the security concerns for laptop computers are also valid for mobile devices.

Smart Systems

A good method of keeping unauthorized users out of your network is to use an access control list (ACL) that screens out all attempts to gain access to your network by wireless network interface cards that do not have a preprogrammed, recognized MAC address.

These systems must evolve to the point where the ACL is dynamically programmed with rules. For example, if there is a spike in network activity during non-peak hours, the computer can identify this trend as potential hacking activity. If a mobile computing device is stolen (or used without the user's knowledge), how is an administrator supposed to know to remove the MAC address of that machine from the ACL? The computer can flag this type of "off-hours" activity as suspicious, so that you can monitor any incoming connections from this machine and deactivate it automatically if it attempts to access secured wireless network resources.

Scrambled Data

If a hacker can't get in the front door by logging into an open wireless network, he will try the back door—attempting to eavesdrop on your wireless network traffic or unencrypted data stream in the hope that your password or other vital information will be transmitted in clear text across the WLAN.

WEP will not provide you with the privacy you expect from encrypting your data. Hackers have found many of its vulnerabilities; therefore you must plan proactively to scramble or encrypt your data using either PKI or VPN mechanisms. In order to ensure that you maintain your privacy, no connection should ever be made to your wireless network unless you have the most sophisticated level of encryption possible.

OS Platform Evolution

It seems that all the major operating systems are always being upgraded with new features and functionality, and this is especially true of capabilities dealing with the automatic recognition of wireless networking cards. An operating system has its own series of vulnerabilities that can be compromised more easily from a wireless network than from wired one. As far as the computer is concerned, the WLAN is indistinguishable from the LAN. A hacker can probe your WLAN for a specific operating system, determine its vulnerabilities, and mount an attack based on acquiring internal network access to control both your host computer and your intranet.

Your best solution involves staying up to date with all the service packs, security fixes, and necessary settings for each vendor's operating system in an attempt to deal with new attacks as they are created. If you can defend against an unauthorized exploit into your computer, you have a reasonable chance of preventing a hacker from destroying the integrity of your online resources.

Windows XP Security

Windows XP has already evolved to incorporate Service Pack 1, as of late 2002. The number of security updates and fixes, however, will be a fact of life due to the increasing number of hacker exploits against this operating system. As these types of attacks grow in severity, it will be imperative that you keep your Windows XP platform constantly updated with all the new fixes as they come out. One way you can do this is to allow Windows to "automatically update and install" all of these fixes as they become available. One of the nice features of Windows is that it is more than willing to do this for you on a constant basis.

Windows XP and Windows 2000 (to some extent) automatically recognize and configure wireless network interface cards to be used on any network within range of your workstation. Windows will allow you to specify the SSID of a station you want to connect to or will allow you to browse the network to find a station that has a strong signal in your area.

Windows is already set up to deal with Wi-Fi cards, so it is imperative that you check your local area network connection to make certain what access points are in your immediate area, and that you don't accidentally roam out of your preferred network to a hacker network. As Windows evolves, more safeguards will have to be put into place to warn the user if the network connection unexpectedly changes or if an interference pattern is degrading the signal, signifying an attempt to break into your wireless workstation.

Macintosh OS X

Macintosh OS X has fully integrated support for its own Airport card. Airport cards are really 802.11b wireless networking interface cards, fully compatible with other brands of WLANs for the PC or Linux. These cards may have a different name, but they have the exact same problems as described in the Windows world. Since Airport is really just another method of 802.11b WLAN, security is still a major problem.

Airport wireless networks must evolve to deal with new security threats just as much as Windows did. Mac OS X has integrated support for automatic updates, and Apple is also very good about putting out automatic security updates to ensure that your operating system is protected against new threats that come out all the time.

Macintosh automatic update is somewhat less invasive than its Microsoft counterpart. The operating system is fully configurable to check on a schedule that you set to see if updates are available. You have the option to deny these updates if you choose. Just like Windows, automatic update can even be turned off, but you must actually change the settings to disable it.

The Mac OS X operating system does offer the same type of stability and security that you would find in other UNIX-based systems, so this platform will most likely evolve to implement more security measures to authenticate users trying to log into the network and make sure that they really are who they say.

Palm and PocketPC

Both these mobile computing devices (commonly referred to as PDAs) have evolved to offer limited integrated support for 802.11b or Bluetooth wireless network interface cards. However, like their more powerful computing brothers, they often lack the ability to offer wireless authentication. Most users don't realize this, or know how to enable encryption for these devices to protect the data transmission from the handheld unit to the wireless server.

Linux

Many Linux distributions are evolving to support several 802.11b network cards right out of the box. SuSE and RedHat are examples of two major distributions that support several WLAN NIC cards, most notably the Orinoco WaveLAN card. The driver is built right into the operating system, and it is easily enabled by editing one of the configuration files.

Linux is becoming a very popular networking platform for end-users as well as for servers. Therefore, wireless networking capabilities will become more of a mainstay in the future. As these operating systems evolve, it will be essential to pay extra attention to your wireless networking security and make certain you keep extensive logs of all network activity so that you can identify any possible hacker intrusions or eavesdropping into your WLAN.

Lindows OS

Lindows OS is really trying hard to take all the best features of Linux and empower its platform to have the ability to run Windows applications. This platform is evolving to support wireless networking capabilities natively. Security will be critical to this operating system because it is a newcomer to the OS world.

Preventing Network Intrusion Attempts

There are two ways in which to prevent network intrusion attempts on your WLAN (Figure 18.5).

1. Use an automated intrusion detection system that uses a form of fuzzy logic to detect and report any possible hacking type of activity. These types of systems are mostly automated and use a preprogrammed set of attack signatures to identify hackers attempting to gain access to your wired or wireless network. If a hacker does try to gain unauthorized access, a computer program monitoring network activity will notify the administrator if any suspicious activity occurs. A good example of this type of intrusion detection system can be found at intrusion.com.
2. Use of a manual intrusion detection system that is staffed by actual people. A company will put a special device inside your internal network that securely transmits network logs and activity to the managed service company, using actual people to look for suspicious activity and inform your administrators if you are experiencing hacking activity. A good example of a company that actually uses

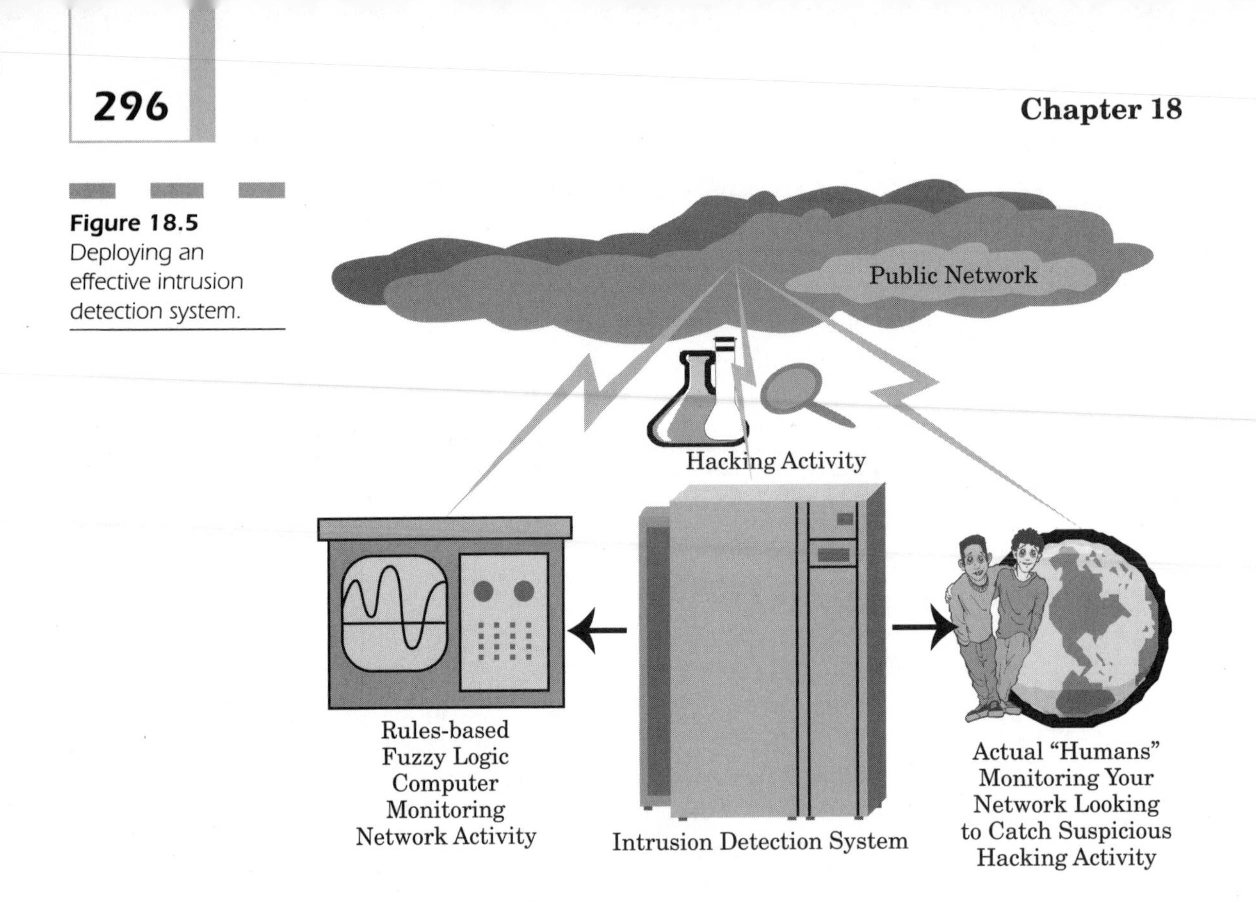

Figure 18.5 Deploying an effective intrusion detection system.

experienced security professionals to monitor your systems for suspicious hacking activity is Counterpane.com.

Network Servers

Network appliances or devices serving up information on the net without being tied to a specific computer server have become very useful over the last few years. Both file and print servers have become useful because if a hacker breaks into them, they only destroy the box, instead of a main server. These devices will grow to offer more functionality with greater security to deal with the barrage of new security threats that occur on a daily basis.

File Servers

File server network devices are often put on the Internet to share files without the risk of having a hacker break into an internal file server that has mission-critical files. These types of devices usually act as "honeypots" to lure a hacker into a system to catch him in the act of trying to gain access. The truth is that you should never put a file server on the Internet that you can't afford to lose.

Printer Servers

Printer servers offer a great deal of convenience, allowing users from virtually "anywhere" to print to any networked printer in your organization. Unfortunately, a hacker who acquires knowledge of your print server can do damage, like making every printer in your organization print non-stop, exhausting all your ink and paper supply.

This type of threat won't go away, but enhancing printer security with an authentication method is an excellent way to ensure that only authorized users will print to your printer server in the future. Simple methods of maintaining access control are often sufficient to help you secure network resources.

Conclusion: The Future of Wireless Networking

While this book demonstrates all the security vulnerabilities of your WLAN, it is clear that because there are so many advantages to using a WLAN in your corporate environment, these devices will not disappear any time soon from the IT landscape within your organization.

While you can never expect to provide 100 percent security for your WLAN, you can take the simple precautions outlined in this book to look for potential vulnerabilities, plug those holes, and prevent hackers from corrupting your resources. If a hacker does break into your network, keeping accurate logs is an excellent way of tracing that network activity so that you can block any future attempts.

Above all, make certain you have good security professionals monitoring your network logs and real-time network activity for any potential problems. Hackers love to try to hack into your network late at night on off hours, or during weekends when very few people are in your corporate facilities. By examining the network logs, it is not only possible to detect spikes in abnormal network usage activity, but you can look for low-level hacking activity, where a hacker attempts to guess at your settings a little each time so as not to trigger any possible alarms you might have configured to detect items such as a distributed denial of service attack on your systems.

Wireless LANs will undoubtedly improve greatly in terms of speed, usability, and security as time goes on. Authentication and PKI mechanisms are only the beginning of locking down your WLAN so that you can control access to any of your networking resources.

For the most part, an ounce of prevention is all you need to prevent damage to your wireless network before it starts. Keep an eye out for suspicious activity and make certain you inform your users to stay vigilant about who has access to your network and what rules you have in place through your security policy so that only specified users have access to selective resources. If you monitor your network and watch all wireless connections, you can be certain that you can provide sufficient security to provide a dedicated wireless network and stay problem free.

INDEX

Note: Boldface numbers indicate illustrations.

ABOUT THE AUTHOR

MR. STEWART S. MILLER has more than a decade of highly specialized technical security and privacy expertise. He has published 11 books in the computer field and over 1000 feature articles. Miller is the country's leading IT security and efficiency management expert. Known best as an executive senior consultant, Stewart has created market analysis/research for hundreds of leading Fortune 500 companies. Stewart has worked with major organizations including IBM and Ernst & Young; he is very well-known for his expertise with complex enterprise systems including SAP, J.D. Edwards, Baan, and PeopleSoft. He has demonstrated his leadership and communication skills as the key-note lecturer for the IBM/SAP Partnership, and literally wrote the book on SAP R/3 Certification. Mr. Miller is known to be "the" industry leader as an efficiency expert in both science and technology because he has collectively saved his clients and users of his materials hundreds of millions of dollars. He is also an IBM Certified IT Security Consultant, charter member of the National Association of Science Writers, and has certifications in every module of SAP and PeopleSoft.